Progress in IS

Progress in IS encompasses the various areas of Information Systems in theory and practice, presenting cutting-edge advances in the field. It is aimed especially at researchers, doctoral students, and advanced practitioners. The series features both research monographs, edited volumes, and conference proceedings that make substantial contributions to our state of knowledge and handbooks and other edited volumes, in which a team of experts is organized by one or more leading authorities to write individual chapters on various aspects of the topic. Individual volumes in this series are supported by a minimum of two external reviews.

Andi Fitriah Abdul Kadir
Arash Habibi Lashkari
Mahdi Daghmehchi Firoozjaei

Understanding Cybersecurity on Smartphones

Challenges, Strategies, and Trends

Andi Fitriah Abdul Kadir
Centre of Excellence for Cybersecurity
(CoExCyS)
International Islamic University Malaysia
Wilayah Persekutuan, Kuala Lumpur,
Malaysia

Arash Habibi Lashkari
Behaviour-Centric Cybersecurity Center
(BCCC), School of Information Technology
York University
Toronto, ON, Canada

Mahdi Daghmehchi Firoozjaei
MacEwan University
Edmonton, AB, Canada

ISSN 2196-8705 ISSN 2196-8713 (electronic)
Progress in IS
ISBN 978-3-031-48864-1 ISBN 978-3-031-48865-8 (eBook)
https://doi.org/10.1007/978-3-031-48865-8

© The Editor(s) (if applicable) and The Author(s), under exclusive license to Springer Nature Switzerland AG 2024

This work is subject to copyright. All rights are solely and exclusively licensed by the Publisher, whether the whole or part of the material is concerned, specifically the rights of translation, reprinting, reuse of illustrations, recitation, broadcasting, reproduction on microfilms or in any other physical way, and transmission or information storage and retrieval, electronic adaptation, computer software, or by similar or dissimilar methodology now known or hereafter developed.

The use of general descriptive names, registered names, trademarks, service marks, etc. in this publication does not imply, even in the absence of a specific statement, that such names are exempt from the relevant protective laws and regulations and therefore free for general use.

The publisher, the authors, and the editors are safe to assume that the advice and information in this book are believed to be true and accurate at the date of publication. Neither the publisher nor the authors or the editors give a warranty, expressed or implied, with respect to the material contained herein or for any errors or omissions that may have been made. The publisher remains neutral with regard to jurisdictional claims in published maps and institutional affiliations.

This Springer imprint is published by the registered company Springer Nature Switzerland AG
The registered company address is: Gewerbestrasse 11, 6330 Cham, Switzerland

Paper in this product is recyclable

Preface

This book is part of the comprehensive Understanding Cybersecurity Series (UCS) research program, aiming to provide diverse cybersecurity resources for researchers and readers from various backgrounds. In 2020, the team released the first online article series, "Understanding Canadian Cybersecurity Laws," which received recognition and was awarded the Gold Medal for Best Blog Column in the Business Division at the 2020 Canadian Online Publishing Awards. Building on this success, the team published the first book, *Understanding Cybersecurity Law and Digital Privacy: A Common Law Perspective*, in 2021 through Springer Nature Switzerland AG.

Continuing their research efforts, the team launched the second article series in 2021, titled "Understanding Cybersecurity Management for FinTech (UCMF)," accompanied by the publication of the related book *Understanding Cybersecurity Management in FinTech: Challenges, Strategies, and Trends*. This book highlights the significance of cybersecurity in financial institutions by showcasing recent cyber breaches, attacks, and financial losses.

Starting in 2022, the UCS team embarked on the third online series, "Understanding Current Cybersecurity Challenges in Law," addressing emerging trends and critical legal issues concerning cybersecurity globally. This series, consisting of six parts, explores digital jurisdictional authority and user-generated digital content ownership. The series is complemented by the publication of the third book, *Understanding Cybersecurity Law in Data Sovereignty and Digital Governance: An Overview from a Legal Perspective*, which offers an in-depth understanding of current cybersecurity challenges and their legal implications.

Simultaneously, the team also worked on the fourth book, *Understanding Cybersecurity Management in Decentralized Finance: Challenges, Strategies, and Trends*. This book comprehensively reviews cybersecurity in blockchain technologies, analyzing platforms like Ethereum, Binance Smart Chain, Solana, Cardano, Avalanche, and Polygon. It explores cybersecurity issues in smart contracts, and related blogs are currently being published through the ITWorldCanada website.

In addition to these endeavors, the team has also focused on understanding cyber threats and adversaries on smartphones, examining cybersecurity threats, vulnerabilities, and risk management. The main objective of this book is to enhance readers' understanding of the evolving cyber threat landscape, encompassing various threat categories and vulnerabilities associated with different smartphone operating systems. The book offers practical solutions for securing and protecting smartphones while raising awareness of the importance of smartphone security.

Wilayah Persekutuan, Kuala Lumpur, Malaysia	Andi Fitriah Abdul Kadir
Toronto, ON, Canada	Arash Habibi Lashkari
Edmonton, AB, Canada	Mahdi Daghmehchi Firoozjaei

Acknowledgments

For all of those fighting for Women, Life, and Freedom.

About the Authors

Andi Fitriah Abdul Kadir is Assistant Professor at the Department of Computer Science at International Islamic University Malaysia (IIUM), Malaysia, and an alumna of the Canadian Institute for Cybersecurity (CIC) at the University of New Brunswick, Canada. She received the IIUM Academic Excellence Award and the Tunku Abdul Rahman Award. She also received several awards from international academic conferences, including the Gold Medal, Best Poster, and Best Paper awards. She works closely with industry and the government, focusing on research and development (R&D) projects. One of her research projects on privacy engineering is part of the 12th Malaysia Plan (RMK12), a strategy for allocating Malaysia's national budget to all economic sectors, including cybersecurity. Her current research focus is computer forensics, network security, malware analysis, machine learning, and privacy engineering.

Arash Habibi Lashkari is Canada Research Chair (CRC) in Cybersecurity. He is a senior member of IEEE and an associate professor at York University. He is the author of ten published books and more than 110 academic articles on various cybersecurity-related topics. He is the co-author of the national award-winning article series "Understanding Canadian Cybersecurity Laws," which was recently recognized with a Gold Medal at the 2020 Canadian Online Publishing Awards. Dr. Lashkari has over 26 years of teaching experience spanning several international universities, has received 15 awards at international computer security competitions—including three gold awards—and was recognized as one of Canada's Top 150 Researchers for 2017. Building on over two decades of concurrent industrial and development experience in network, software, and computer security, Dr. Lashkari's current work involves the development of vulnerability detection technology to protect network systems against cyberattacks. He simultaneously supervises multiple research and development teams working on several projects related to network traffic analysis, malware analysis, Honeynet, and threat hunting.

Mahdi Daghmehchi Firoozjaei is Assistant Professor in the Department of Computer Science at MacEwan University, Canada. Prior to this, he was Assistant Professor at the University of Windsor, Canada, and a postdoctoral research fellow at the University of New Brunswick, Canada. He received a BSc degree in telecommunication engineering and an MSc degree in cryptology. Dr. Firoozjaei did his PhD in computer engineering at Sungkyunkwan University, Korea, in 2018. He has 10 years of industry experience in telecommunication systems and cybersecurity. He worked for the Telecommunication Company of Iran (TCI), Babol, Iran, as a senior engineer from 2006 to 2014 and was a research fellow at the Canadian Institute for Cybersecurity (CIC), Canada, from 2018 to 2021. Dr. Firoozjaei led R&D projects in the operational technology (OT) forensics for Siemens Canada and DNS firewall for CIRA. His research interest focuses on persistent malware analysis, digital forensics, network security, privacy-preserving, and blockchain.

Contents

1	**Introduction**	1
	1.1 Smartphone History	1
	1.2 Smartphone Market Share Evolution	3
	1.3 Security and Privacy	4
	1.3.1 Application Stores	5
	1.3.2 Security Principles (CIAAA)	6
	1.3.3 Privacy	8
	1.3.4 Vulnerabilities	9
	1.4 Cybersecurity Challenges and Emerging Trends in Smartphones	10
	1.4.1 Adversarial Techniques	12
	1.4.2 Attack Types and Impacts	16
	1.4.3 Malware-on-the-Go	17
	1.5 Chapter Summary	20
	References	21
2	**Android Operating System**	25
	2.1 Learning Basics: Android History	25
	2.2 Getting into Cybersecurity: Android Vulnerabilities and Risks	29
	2.3 Adversarial Techniques	32
	2.4 Dissecting Malware: Types of Android Malware	34
	2.5 Mitigating Attacks: The Current Solutions	36
	2.6 Utilizing Android Services: The Trend Now	38
	2.6.1 Android for Industry 4.0	38
	2.6.2 Android Things	38
	2.6.3 Android Enterprise	39
	2.7 Chapter Summary	40
	References	40
3	**iPhone Operating System (iOS)**	43
	3.1 Learning Basics: iOS History	43
	3.2 Getting into Cybersecurity: iOS Vulnerabilities and Risks	44
	3.3 Adversarial Techniques	46

	3.4	Dissecting Malware: Types of iOS Malware	46
	3.5	Mitigating iOS Attacks: The Current Solutions	51
	3.6	Utilizing iOS Services: The Trend Now	52
		3.6.1 Wearable Technology	52
		3.6.2 Mobile Wallets (Use of Apple Pay)	52
		3.6.3 Augmented Reality (AR) and Virtual Reality (VR)	52
		3.6.4 Voice Assistants	52
		3.6.5 iOS App Security	53
		3.6.6 iOS HomeKit	53
	3.7	Chapter Summary	53
	References	54
4	**Windows Operating System**	**57**
	4.1	Learning Basics: Windows OS History	57
	4.2	Getting into Cybersecurity: Windows Vulnerabilities and Risks	60
	4.3	Adversarial Techniques	62
	4.4	Dissecting Malware: Types of Windows OS Malware	63
	4.5	Mitigating Windows Attacks: The Current Solutions	64
	4.6	Utilizing Windows Mobile Services: The Trend Now	66
	4.7	Chapter Summary	67
	References	68
5	**Other Operating Systems**	**71**
	5.1	Learning Basics: The History	71
		5.1.1 Symbian OS	72
		5.1.2 Tizen OS History	74
		5.1.3 Sailfish OS History	75
		5.1.4 Ubuntu Touch OS History	76
		5.1.5 KaiOS History	77
		5.1.6 Sirin OS History	78
		5.1.7 HarmonyOS History	78
	5.2	Getting into Cybersecurity: The Vulnerabilities and Risks	80
	5.3	Dissecting Malware: The Type of Malware	83
	5.4	Mitigating Attacks: The Current Solutions	84
	5.5	Utilizing Services: The Trend Now	85
	5.6	Chapter Summary	86
	References	86
6	**Mobile Application Security**	**89**
	6.1	Application Security Threats	89
	6.2	Application Vulnerabilities	91
	6.3	Information Sensitivity	92
	6.4	Semantic Security	93
	6.5	Infrastructure-Related Security	93
	6.6	Privacy Awareness	94
	6.7	The Best Practices for Mobile Users on App Security	94

	6.8	The Best Practices for Smartphone App Developers	96
	6.9	Chapter Summary	100
	References	101
7	**The Best Security Practices**	103
	7.1	Chapter Summary	106
	References	106
8	**Conclusion**	109
	References	113

Chapter 1
Introduction

The smartphone is without a doubt one of the greatest inventions of modern humanity and, today, is the most extensively used electronic gadget in the world. The development of the smartphone revolutionized our modern communication technology and provided us with the ability to communicate effectively and instantaneously over long distances and around the world. This chapter provides readers with the historical context in the development of the modern smartphone and highlights some of the cybersecurity-related concerns relevant to the smartphone and its various applications (apps).

1.1 Smartphone History

The smartphone, as we now know it, was originally launched to the broad consumer market back in 2007 by Steve Jobs and Apple. However, if we dig into the history, the smartphone has been around since 1993. The primary differences between the smartphones of today and the earlier versions of smartphones are that early smartphones were designed for corporate customers, were mainly utilized as enterprise devices, and were also prohibitively expensive for the vast majority of consumers [1, 2].

To progress to the point that it has reached today, mobile technology has undergone many changes. In the technology industry, we refer to the clusters of technological evolutions as "generations," which refers to the maturity and the capabilities of the real cellular networks. We can take a trip down memory lane [3], as portrayed in Fig. 1.1, to observe when each groundbreaking development occurred:

- **1970s: The Brick Era**: On April 3, 1973, Motorola employee, Martin Cooper, made the first public mobile phone call in New York using a Motorola DynaTAC. The phone was called a "brick" due to its size and weight.

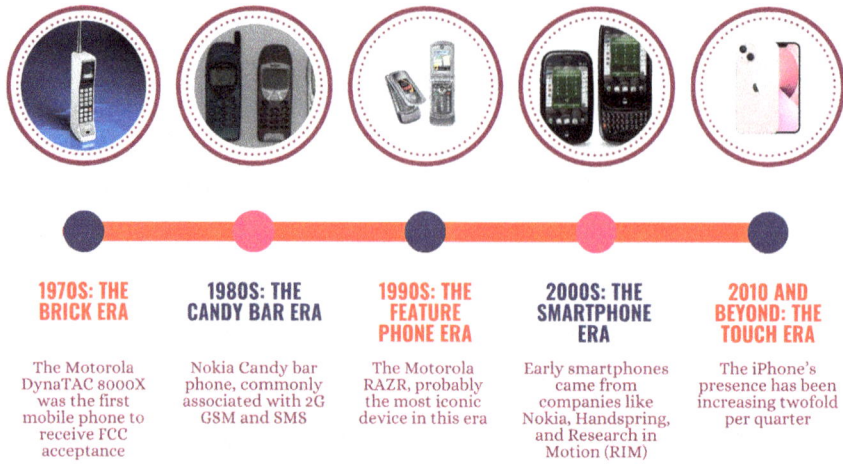

Fig. 1.1 The eras of mobile phones: from "bricks" to "clicks"

- **1980s: The Candy Bar Era:** This is one of the most significant advancements in mobile technology. The majority of mobile devices in this era have a long, thin, rectangular form factor, which is referred to as a "candy bar."
- **1990s: The Feature Phone Era**: In 1991, Finland became the first country to switch to second-generation (2G) technology. A year later, IBM invented the first smartphone, and it was available for purchase in 1994. Its name was the Simon Personal Communicator (SPC).
- **2000s: The Smartphone Era**: The smartphone was not connected to a real 3G network until the year 2000. In other words, a mobile communications standard was created to allow wireless Internet access to portable electronic devices. In 2009, the long-term evolution (LTE) 4G standard was initially deployed in Stockholm, Sweden and Oslo, Norway.
- **2010 and beyond: The Touch Era:** Smartphone interfaces shifted away from physical keyboards and keypads to huge finger-operated capacitive touchscreens in the late 2000s and early 2010s. By the mid-2000s, most high-end cell phones had built-in digital cameras. For the record, the iPhone 5S, released in September 2013, was the first smartphone on a major US carrier to include this technology since the Atrix. In March 2019, three South Korean carriers, KT, LG Uplus, and SK Telecom, began commercial 5G services.

1.2 Smartphone Market Share Evolution

Many people, yourself included, have probably engaged in a "PC versus Mac" debate. Ultimately, this debate boils down to individual personal preference; what you like and what you want to do with your system. There is a similar dispute regarding the merits of alternative mobile operating systems (OS) in the world of smartphones. The most popular mobile operating systems—including iOS, Android, Blackberry OS, and Windows Mobile—along with major smartphone vendors—such as Apple, Samsung, HTC, Motorola, Nokia, LG, and Sony Ericsson—are focusing on bringing options to both the operating systems and the devices that will provide exciting features to enterprise and general consumers alike [4].

Notably, the Android sector played a huge role in this development, as it allowed all varieties of vendors to build devices using unprecedented open-source Android technology. As a result, the Android OS market share is expanding every year, as demonstrated in Fig. 1.2.

The touchscreen smartphone revolution began in 2010 and has had a significant impact on basic feature phone sales, with smartphone sales increasing from 139 million units in 2008 to 1.54 billion units in 2021. As a result of the impacts of the coronavirus (COVID-19) pandemic, smartphone sales briefly fell to 1.38 billion units in 2020 but recovered the following year. In this shift toward smartphones,

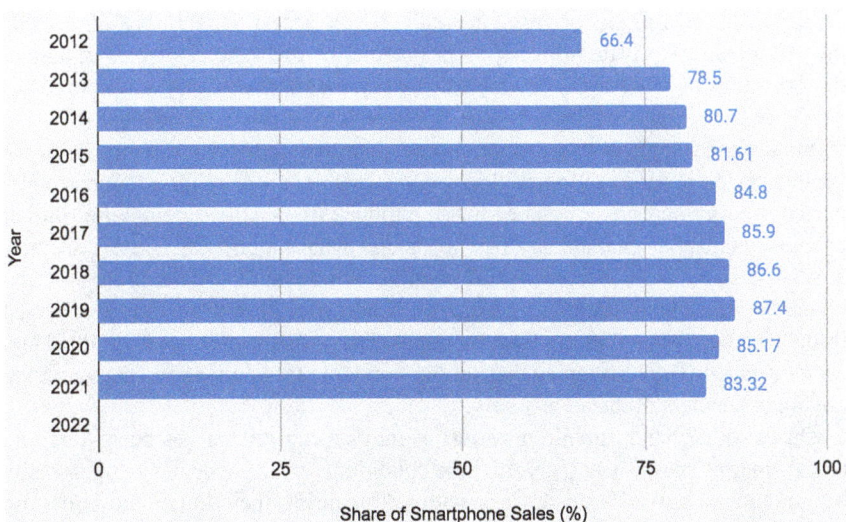

Fig. 1.2 Android OS market share (from 2011 until 2022) [Data are collected from Statista (2022)]

Apple, Samsung, and, more recently, Xiaomi were the great winners, leaving Nokia behind. In just 3 years, Nokia's net revenues dropped by about 30 billion euros. It has been speculated that one of the significant reasons behind Nokia's significant drop in revenue is the company's focus on hardware rather than on software standards.

1.3 Security and Privacy

Nowadays, smartphones are essential gadgets in our lives. Undoubtedly, no other tool in modern history has revolutionized our lives as much as smartphones have. These handheld computers have replaced many devices, e.g., cameras, GPS, calculators, and address books. Due to the advance in smartphone's technologies, these devices are doing almost everything a desktop computer can do. Most of our daily electronic tasks, e.g., Internet browsing, social media networks, banking, online gaming, online shopping, and professional activities, are done through smartphone apps. These apps' increasing use and integration into our daily lives provide opportunities for innovation and community benefits [5]. Despite these benefits, smartphones are a distraction and allow the user to rely on the device too much for daily activities. The problem is that many users overshare their information on these channels. There are so many application stores (authorized or unauthorized) that share smartphone applications and gadgets (free or licensed) for different purposes. Aside from personal information, applications may need permission to access various features on the mobile device. For example, if a user wants to take a picture using Instagram, the app will need permission to use the device's camera. There is a massive number of permissions an app could request, but not all permissions are the same. Wrong or risky permission could provide access to data or resources that involve the user's private information or potentially affect the user's stored data or the operation of other apps. Examples of risky permissions include access to the user's location, contacts, SMS messages, phone logs, camera, or calendar [6].

Smartphone applications analyze users' data and retain their information for beneficiary purposes (e.g., targeted advertisements or user profiling). It opens unexpected security and privacy issues for smartphone users. Cloud computing further enables data to be obtained. It does not matter what mobile OS is used, security attacks and privacy breaches are still possible by transmitting data through smartphone applications in iPhone, Android, or Windows phones. Data that is collected through those applications are used for targeted advertising after it is assembled into users' profiles. The companies that set the standards for application data gathering are big contenders in the ad business.

On the other hand, mobile network operators can see what has been stored on users' smartphones if necessary. In these conditions, smartphone devices can easily be turned into surveillance devices without impairing their functions. Users are profiled by tracking their devices by hackers, the government, or cloud/network service providers [7].

1.3 Security and Privacy

1.3.1 Application Stores

A smartphone application, also known as mobile app, is a software to run a kind of application on a smartphone. Initial mobile apps provided general-purpose information and information services on the global network, e.g., email, calendar, stock market, and weather information. The advent of new online services and high demand of smartphone users, along with the ability to develop mobile apps extend it into other categories, such as mobile games, factory automation, banking, health services, and so on. The explosion in the number and variety of applications has developed into large and diverse areas. Nowadays, many services need the help of mobile app technology such as identifying users' locations for tracking, monitoring users' behavior, purchasing tickets, and even medical services [8].

To distribute mobile apps, there are some digital distribution platforms called mobile application stores (app stores). Depending on the smartphone's OSes, apps are designed and categorized by an app store. Typically, an app store organizes mobile apps based on their OSes, the functionality (e.g., productivity, games, or multimedia), devices which apps are designed for. App stores provide a digital store in which users can search through different categories, view apps' information, and digitally purchase [9]. While Google Play,[1] Amazon AppStore,[2] GetJar,[3] and Aptoide [4] are top Android app stores, App Store,[5] iTune,[6] and Appland[7] are top iOS app stores [10], and Windows Apps[8] for Windows Phones. Beside these large digital stores, some purpose-designed apps are directly offered by the developers (e.g., third-party apps) or service providers, such as mobile device manufacturers (e.g., Galaxy Store[9] and BlackBerry[10]), car companies (e.g., myAudi[11]), health service providers, insurance/banking companies, and so on.

App stores, which their owners curate, accept a submitted app after an approval process. An inspection process typically includes quality control, censorship, and security checks. Some app stores provide users' reviews, allow for grading, and reflect the feedback to developers, including the number of installations and practical issues, e.g., latency and installation errors. Despite high-level security checks employed by app stores, the open-source software submission nature allows malicious developers to develop malicious apps. Most malware apps are hidden under

---

[1] https://play.google.com/
[2] https://www.amazon.com/gp/mas/get/android/
[3] https://www.getjar.com/
[4] https://en.aptoide.com/
[5] https://www.apple.com/ca/app-store/
[6] https://www.apple.com/ca/itunes/
[7] https://www.applandinc.com/
[8] https://apps.microsoft.com/store/apps
[9] https://www.samsung.com/ca/apps/galaxy-store/
[10] https://www.blackberry.com/us/en
[11] https://my.audi.com/

the cover of regular mobile apps developed by third-party developers and are shared on unauthorized online markets [8]. Downloading mobile apps from unauthorized app stores leads to security threats of mobile malware apps. It is difficult for users to identify the malware against the actual apps. Employing inadequate security check mechanisms in the app store's use leads to sharing vulnerable mobile apps.

1.3.2 Security Principles (CIAAA)

To establish a robust and secure system, it is essential to ensure the implementation of five fundamental security principles. These principles are confidentiality, integrity, availability, authenticity, and accountability (CIAAA). CIAAA is a crucial factor in securely operating a system. CIAAA helps to construct security strategies and develop policies and controls, while also confirming as a foundational starting point for any unknown use cases, products, and technologies [10]. Based on this, the security principles are segmented into separate critical issues. This differentiation helps pinpoint the different ways to cope with and address each security concern. Accomplishing all standards of CIAAA makes the system's security profile stronger and equips it to handle threat incidents better.

Data Confidentiality

Confidentiality refers to protecting information from any unauthorized access to avoid unauthorized disclosure. Maintaining data confidentiality, including ensuring privacy and data breaches, achieves multiple goals. Usually, confidentiality threat attacks aim to intercept access to the data. Keyloggers and port scanners are examples of these attack mechanisms. Ransomware attacks, for instance, are implemented through access mechanisms. To accomplish confidentiality, access to information must be supervised and controlled to prevent unauthorized access to data. A critical component of preserving confidentiality is making assure that people without proper authorization are stopped from accessing assets.

Data Integrity

Data integrity concepts include processes to ensure a user's data's accuracy, completeness, consistency, and validity. These processes provide the integrity of the data and guarantee that the user has accurate and correct data. The importance of data integrity increases as the user's data volumes continue to grow exponentially. Data integrity can be compromised (e.g., manipulated), while it is stationed, replicated, or transferred. Data integrity should remain intact and unaltered between updates and transmissions. Error-checking methods and validation procedures are typically relied on to ensure the integrity of data that is transferred or reproduced without the intention of alteration [11].

There are different types of integrity, namely physical integrity and logical integrity. Physical integrity includes issues related to storing and retrieving data, primarily the storage devices, memory components, and any associated hardware. Logical integrity depends on factors that affect the correctness or sensibility of data within its respective context. Poor software design, bugs, human error, and malfeasance affect logical integrity. While physical and logical integrity may have distinct definitions, they are frequently interconnected. So, data integrity can be compromised because of various factors, which encompass human errors (such as accidental data manipulation or deletion), transmission errors (including noise, network failures, or incorrect storage destinations), malicious activities (such as malware infections, hacking, or data theft), and inadequate infrastructure configurations (such as weak network setups or insufficient security mechanisms) [11].

Data Availability

Data availability guarantees that the data is available whenever it is required. This means that systems, networks, and applications must function as they should and when. Availability can also be compromised through deliberate sabotage, such as denial-of-service (DoS) attacks or ransomware. Organizations can use redundant networks, servers, and applications to ensure availability. These can be programmed to become available when the primary system has been disrupted or broken. Data availability can be enhanced by staying on top of upgrades to software packages and security systems. In this way, the user makes it less likely for an application to malfunction or for a relatively new threat to infiltrate their system. Backups and full disaster recovery plans also help a company regain availability soon after an unwanted event.

Authenticity

Authenticity ensures that information and communication come from a trusted source. It is divided into user authentication and data authenticity. Data can be authentic if it is probable that it has not been corrupted after its creation. Data authenticity plays a crucial role in ensuring the identity and ownership of data. It involves verifying the source or origin of data and other file transfers by providing proof of identity [12]. This means that a digital object is indeed what it claims to be or what it is claimed to be. The main benefit of data authenticity is ensuring that the message (e.g., email, payment transaction, and digital file) was not corrupted or intercepted during transmission. Through user authentication processes, users can verify their identities by providing specific credentials (e.g., username and password, biometric data, digital signatures, or authentication tokens). This includes protecting against impersonation, spoofing, and other identity fraud and attacks.

Common techniques used to establish authenticity include authentication, digital certificates, and biometric identification. A mobile app can directly do data and user authenticity. It can verify that the original sender has sent a delivered message and has not been altered during the transmission. The message itself can be in clear or encrypted form.

Accountability

Since users have no control over their data at mobile networks or service provider centers, data accountability enables mobile service/app providers to give their users appropriate control and transparency over how their data is handled and used. It brings the confidence that users' data is handled according to their expectations and is protected [13]. The accountability concept helps to guide explanations for data practices that seem to be insufficient. The concept of accountability compels organizations to implement measures that adhere to relevant privacy legal requirements. They should be able to demonstrate the existence and effectiveness of such measures internally and externally upon request. This means that organizations should implement comprehensive data privacy and security programs that cover all aspects of data processing, including collection, use, transfer to third parties, and disposal [14]. Accountability and transparency are closely linked together [15]. Ensuring accountability of data necessitates transparency in how data is handled. The level of transparency exhibited by an organization demonstrates its capabilities in safeguarding privacy protection. In this view, accountable systems can report back to users on how their data is being managed, who has accessed their data, and when and what modifications have been performed on their data. Only by knowing what is going on with their data can users be sure that their data is being used effectively [16].

1.3.3 Privacy

Due to the advance of new technologies, smartphone devices are becoming more intelligent and well-equipped. Through different sensors and on-device data storage, smartphones have access to an array of the most personal information, are constantly running, and are almost always on the person of a mobile user. Smartphones have a variety of sensors, such as GPS, Wi-Fi, Bluetooth, accelerometers, gyroscopes, microphones, and cameras. These devices emit various signals containing unique identifiers, which can be captured for tracking and profiling. Smartphones collect unprecedented amounts of data, which can reveal sensitive and highly personal information. This personal information is shared with authorized or unauthorized providers (e.g., third-party service providers). This leads to privacy issues for data and mobile users, e.g., semantic or location privacy issues. For instance, tracking users' location data shared by a mobile navigation app can seriously threaten users' location privacy [17].

Analyzing users' private information facilitates providers in monitoring mobile users' activities and predicting their behavior. This analysis, also known as user profiling, is commonly used by service/mobile app providers. User profiling includes analytical approaches that lead to predictive maintenance, abnormality, and fault detection used in service/network management processes. While most users show variable behavior, they typically exhibit a certain amount of repetitiveness [18]. A mobile user's behavior is affected by their social, professional, and economic situations and time factors, e.g., being on vacation, work time, or free time. User profiling can be used to predict and balance service purchases and sales.

To avoid a privacy breach, the main mobile app stores, namely Google and Apple, employ technical mechanisms to put conditions on the data that app providers can access when operating on users' smartphones. A privacy breach is the loss of unauthorized access to, or disclosure of, users' personal information. In addition to the app stores, a fundamental control and protection mechanism are needed at the smartphone OS level. Mobile OSes, such as iOS and Android, use permission architecture for apps accessing various mobile devices and data functions. The main purpose of these permissions is to protect user privacy. App sandboxing is a technical feature used by mobile OSes to isolate users' data in one app from other apps and protect user data from unwanted access by other apps. This privacy-by-design technique isolates apps within containers that only hold data that the app generates [17].

1.3.4 Vulnerabilities

Sending and receiving confidential information on mobile devices is risky due to untrusted network provider. In the cellular network communications, all data generated by the mobile is handled through the service provider's network. Voice calls, SMS/MMS, and Internet data are sent to the cellular network provider to deliver to the destinations. Network provider may retain user's data, share with the third parties, or monitor data traffic for management purposes. On the other hand, users often use unsecured Wi-Fi connections, e.g., public places, with no attention to potential security threats. Unfortunately, hackers can easily break in. Mobile apps often do not encrypt data before transmissions, making that data vulnerable while it is in transit [19]. Man-in-the-middle (MITM) attack, data tampering, and data injection are possible with an untrusted or a compromised network provider [20].

Insecure data storage in the smartphone device or service providers' datasets are a potential vulnerability. When the provider is compromised or the smartphone device is infected by the malware, data breach is possible in the case of insecure storage. Weak authentication mechanisms, employed on smartphones, open a vulnerability for hackers. Many mobile apps rely on password-based authentication, as a single factor authentication. This exposes users to different threats, including stolen credentials and automated brute force attacks. Improper session handling has common vulnerabilities when using internet applications over mobile devices [21].

Smartphone device built-in sensors, known as background sensors, provide frequent measures of physical quantities in an unobtrusive and transparent way. However, these data can be easily utilized to extract sensitive information of the user such as gender, age, emotion, ethnic group, and so on [22]. Analyzing this information leads to profile mobile users, which is an important privacy issue. Although user profiling has several beneficiaries for service/network providers, it can easily compromise mobile users' security and privacy. Credential extraction by data analysis attack, authentication compromising by replay attack and masquerading, and privacy breaches are possible security and privacy issues with the user profiling [23–26].

1.4 Cybersecurity Challenges and Emerging Trends in Smartphones

You might be surprised to learn about the hidden security risks that lay hidden inside your trusted mobile device. Indeed, although the smartphone has brought great improvements to our lives today, it has also presented significant challenges to cybersecurity. This section describes the most significant "information security" threats on smartphones, which include the following:

- **Data leaks**: The unintended or unlawful flow of personal information from a mobile phone to the Internet is known as a data breach or a data leak. Data breaches or data leaks are mostly caused by malicious, or even authorized, mobile apps. According to studies [27], the most frequently exposed data falls under the category of Personally Identifiable Information (PII), which includes information such as usernames, passwords, and credit card details.
- *How to mitigate?* Smartphone users should only grant an app the data access rights that are absolutely necessary in order for the app to function properly, and users should be wary of apps that ask for more data access than is necessary for the app to function.
- **Phishing:** Phishing is the most common incident that today's mobile users encounter. Mobile users are particularly vulnerable because they access their emails in real-time and tend to read their emails on the fly. Furthermore, the smaller screen size makes it more difficult for smartphone users to spot URLs that appear to be suspicious, in contrast to the visibility we are used to seeing on a larger laptop or desktop computer screen.
- *How to mitigate?* Users can take care to avoid accessing unexpected email links, only download programs from trusted app stores, install antivirus software, and use browsers with security features to avoid falling for a mobile phishing scam.
- **Malware:** These programs, also known as malicious software, take advantage of faults in operating systems in order to steal data, modify device configurations to

download more malicious software, display pop-up ads or send a torrent of premium SMS texts for monetization, and even to cripple devices for short period spyware, ransomware, banking malware, scareware, and adware are some of the many known subtypes of malware families.
- *How to mitigate?* Users should use a VPN, along with a vulnerability scanner, and should only download apps from legitimate app stores. Users should avoid "jailbreaking" their phone, which involves changing the phone's settings to provide unlimited access to the file system, as this increases the vulnerability of the phone, making it much more likely to be victimized by malware.
- **Insecure Wi-Fi networks:** For cybercriminals, Wi-Fi hotspots offer a tempting attack vector for extracting data from mobile devices, including smartphones. Attackers utilize well-known public Wi-Fi names Service Set IDentifier (SSIDs) to fool smartphone users into connecting to their fake networks, opening them up for exploitation.
- *How to mitigate?* Users should try to stay away from connecting to free Wi-Fi networks, in general, and should never use free Wi-Fi networks as a means to access sensitive and important data such as their online banking logins.
- **Lock screen configuration:** Even though the lock screen is one of the most basic and important smartphone security features, many users fail to use it for its intended purpose. When the screen lock feature is not being utilized properly, another individual can very quickly access important user data, personal information, emails, and apps if the mobile phone becomes lost, gets stolen, or is left unattended.
- *How to mitigate?* Users should configure a strong password-protected lock screen and enable multifactor authentication (MFA). With MFA, the users must supply two or more verification factors in order to gain access to the internal system of the phone.
- **Outdated operating systems (OS):** By delaying security patches or OS updates, users unnecessarily put their mobile devices at risk. By exploiting obsolete devices and apps, cybercriminals can fairly easily extract private information including credentials, banking information, and social security numbers stored through the smartphone.
- *How to mitigate?* Users should update their OS and applications as soon as possible. Manufacturers offer OS updates on a regular basis, which include not just performance enhancements but also critical security patches for the known flaws that may be actively exploited.
- **Improper application permission:** Users have access to millions of mobile apps, and while some are safe and secure, many are not. This includes apps that can be purchased or freely downloaded from the Google Play and Apple App stores. These apps could be hacked, exposing critical personal information to third parties.
- *How to mitigate?* Users must pay close attention before authorizing any permissions to apps. There is always a risk of compromise, no matter where these apps are available, even though the more legitimate sources.

1.4.1 Adversarial Techniques

Adversaries use a variety of different techniques to compromise users' mobiles. Due to different vulnerabilities because of misconfiguration, limited security protection, or OS's native functions, adversaries achieve tactical goals by performing different techniques. Generally, these techniques involve device access and network-based effects that can be used by adversaries without accessing to a device.

To completely infect a mobile system, an adversary may go through several phases and use different techniques, which can be categorized as follows [28]:

- Propagation
- Activation
- Carrier
- Execution
- Persistence

At each phase, adversaries use different tactics to conduct their attacks. Depending on the victims' hardware, network traffic, operating system, or applications installed on the smartphone, a variety of adversarial techniques are employed.

- Propagation phase
- In the propagation phase, the targets are created, and the adversary attempts to access the mobile target set and deliver the malware file. For instance, this may occur when a user is compromised by a spear phishing attempt to visit a website that is the source of the malware. Some adversarial techniques that are used at this phase are listed as follows [29]:
  - **Clipboard data stealing**
  - Clipboard manager APIs provide users with the ability to manage clipboard commands, e.g., copy, cut, or paste. By abusing these APIs, adversaries can obtain sensitive information that was previously copied to the smartphone clipboard. A malicious application installed on the smartphone can capture passwords copied and pasted from a password manager application. For instance, GolfSpy malware can hijack an infected Android smartphone and obtain the contents of the clipboard [30]. With cyberespionage capabilities, this type of malware can steal sensitive information from infected devices, including individual device accounts, lists of installed applications, and stored files [31].
  - **Access notifications**
  - Smartphone users regularly receive notifications sent by the network operators, application servers, governmental authorities, etc. at different occasions. Generally, notifications contain some data that has useful information for adversaries, such as the user's status or the one-time authentication codes sent over SMS. Adversaries may remove or dismiss notifications in order to prevent the user from being notified. For instance, Mandrake [32] is an Android espionage malware that can capture all device notifications as well as hide device notifications from the user.

1.4 Cybersecurity Challenges and Emerging Trends in Smartphones

- **Drive-by compromise**
- An adversary compromises a legitimate website to inject a malicious code, such as JavaScript, iFrames, and cross-site scripting. When visited by a user, the scripts automatically execute, typically searching vulnerable versions of the browser and plugins. Upon finding a vulnerable version, the exploit code is then delivered to the browser. This technique was used in INSOMNIA [33] and Pegasus to deliver iOS malware to iOS devices and in Stealth Mango [34] to deliver Android malware to Android devices.
- **File and directory discovery**
- To discover any vulnerability on a victim's smartphone, an adversary may explore and identify available files and directories in order to plan proper invasion attacks and compromise the user. To protect users on Android OSes, Linux file permissions and SELinux policies restrict applications from accessing the system files. Similarly, the security architecture of the iOS restricts any unauthorized access for file and directory discovery. Despite these policies, the contents of the external storage directory are generally visible, which could present concerns if sensitive data is inappropriately stored there. For instance, Desert Scorpion (An Optimized Decision Tree based Android Malware Detection Approach using Machine Learning) [35], an Android surveillanceware, lists files stored on external storage.

- Activation phase
- After successfully propagating malicious codes onto the victim's mobile, the adversary avoids being detected in the activation phase. There are several techniques for defense evasion, including uninstalling/disabling security software or obfuscating data and scripts. Some methods used at this phase are as follows:

  - **Elevation control mechanism**
  - Despite the many benefits of the smartphones' APIs, they can also be abused for malicious purposes. For instance, the administrator's API of the device corporates installed applications and allows the smartphone settings to be reset to the factory defaults. Abusing this API, adversaries can remove the malware artifacts, making it more difficult to uninstall the malware, or making a malware able to programmatically grant itself administrator permissions in order to be installed without user input [28].
  - **Exploitation for privilege escalation**
  - To gain higher levels of access on the smartphone system, an adversary can exploit available vulnerabilities in the mobile OS or in the applications running at higher permissions. Through privilege escalation, the adversary takes advantage of a programming error in an application within the operating system software or a service within the kernel to execute an adversary-controlled code. Agent Smith malware [36], for example, exploits known OS vulnerabilities (e.g., Janus) and replaces existing legitimate applications with malicious versions.

- **Hooking**
- To evade detection, and to hide the presence of malicious artifacts, hooking techniques are often used by adversaries. These techniques can be used to modify the return values or data structures of system APIs and function calls. Android devices are particularly vulnerable to this technique. For instance, Monokle malware [37] is an Android Mobile surveillance ware that hooks itself to appear invisible to the Process Manager.
- **Impair defenses**
- Adversaries may maliciously modify the environmental components of a victim's smartphone in order to hinder or disable the inherent defensive mechanisms (e.g., antivirus mechanism). For instance, Android device administration API can be abused to prevent the smartphone user from uninstalling a malicious application. Mandrake, an Android espionage malware [32], resists being uninstalled by abusing the device administrator permissions until its permissions are eventually revoked. Similarly, Cerberus ransomware disables Google Play Protect in order to prevent its discovery and deletion in the future [38].

- Carrier phase
- If additional code, instructions, or data are needed to execute the malware, carrier or secondary channels are used to contact a remote data source. All activities required for executing the malware (e.g., modifying access control or replacing/deleting files) are done in the carrier phase. The following techniques are used at this phase:
  - **Adversary-in-the-middle**
  - An adversary can position itself between separate networked devices in order to redirect the traffic between them. For instance, by registering a malicious application as a VPN client to a smartphone, the adversary is able to perform follow-up behaviors, e.g., transmitted data manipulation or denial of service attacks. Although registering a VPN client requires the user's consent on both Android and iOS systems, if a privilege escalation is possible, then the adversary can gain access to network traffic as an adversary-in-the-middle [28].
  - **Application layer protocol**
  - To avoid detection or network filtering, adversaries may blend the malicious traffic (e.g., commands to smartphone or exfiltrate data) within the application layer protocol traffic, including into web browsing, transferring files, electronic mail, or DNS traffic. For instance, in the DoubleAgent malware family, which was an advanced Android remote access tool (RAT), [29] FTP and TCP sockets were used to allow for data exfiltration. The malicious surveillanceware traffic was embedded within the traffic of those protocols, between the mobile device and server.
  - **Archive collected data**
  - To hide information collected from a smartphone, adversaries may compress or encrypt data. Data compression allows for the obfuscation of its contents and minimizes the use of network resources. By encrypting the archived or

1.4 Cybersecurity Challenges and Emerging Trends in Smartphones

compressed data, adversaries can avoid detection or make the data exfiltration less conspicuous upon inspection by a defender. Golden Cup [39], GolfSpy [31], and Triada [40] malware are used to encrypt and collect data, and FrozenCell [41] malware is used to compress and encrypt data before exfiltration.

- Execution phase
- After the preparation phases, the malware on the victim's mobile executes its payload. In some cases, additional code or instructions are needed for this to execute. In this case, extra channels are used for the adversary's resources. The adversary may steal data from the infected mobile set by input capturing or credential API hooking [37]. The adversary can leverage and abuse mobile OS's features (e.g., the Native Development Kit (NDK) in Android) to write native functions to execute malicious programs without user permission [38]. Some techniques used at this phase are as follows:

  - **Endpoint denial of service**
  - Denial of service (DoS) attacks are performed to degrade or block the access to services for users. On Android devices (prior to versions 7), Device Administrator access was abused in order to reset the device lock passcode and prevent the device from being unlocked. For instance, Xbot [42] was an Android malware that remotely locked infected devices and then asked for a ransom in order to allow the device to be unlocked again by the user.
  - **Fake traffic generation**
  - To generate fraudulent advertising revenue or fake reviews, adversaries may generate outbound traffic from infected devices. HummingBad or HummingWhale [43] (Information security breaches and precautions on Industry 4.0) is an Android malware that displays fraudulent ads in order to generate revenue. This strategy installs applications that generate fraudulent advertising revenue and collect personal data to be sold along the way.

- Persistence phase
- In the persistence phase, the adversary tries to maintain their foothold after malware execution. Any access, action, or configuration changes that support this foothold are part of this phase [44].

  - **Compromise application executable (persistence)**
  - An adversary may modify an application installed on a user's device in order to carry out malicious tasks. The vulnerabilities of smartphone operating systems can be used to inject malicious codes into legitimate applications. For instance, the Janus vulnerability (CVE-2017-13156) in the Android OS was used to modify the code in applications without affecting their signatures. Adversaries could add extra bytes to APK (application) files and to DEX (executable) files and prepend a malicious DEX file to an APK file, without affecting its signature. The APK file was then accepted by the Android OS as a valid update of a legitimate earlier version of the app [45].

- **Boot/logon initialization scripts**
- In mobile OSes, all the scripts necessary for the system are configured correctly at boot time. These initial scripts are not accessible to the user unless the mobile device has been rooted due to misconfiguration or the third-party customizations. In that case, an attacker is allowed to escalate privileges to root [46]. By this privilege, the adversary executes the malicious scripts at boot or logon initialization to establish persistence. For instance, in Android OldBoot malware [47], the adversary with the escalated privileges modifies the "init" script on the device's boot partition to maintain persistence.

1.4.2 Attack Types and Impacts

Informing and educating smartphone users about potential security risks and how to avoid them is one of the major responsibilities taken on by cybersecurity researchers. This is due to the fact that users of smartphones are often unable to assess the threats or vulnerabilities that exist on their devices, including those that may be found within malicious applications [48].

We can look to Table 1.1 to understand more about the different types of attacks and their impacts. In this example, the attacks are classified according to the four assets of our smartphones: data, application, physical, and network [28].

Table 1.1 Smartphone possible attacks and their impacts

Category	Possible threat	Impact
Data	Information stealing or data theft	Loss of information and data
	Data leakage	
Application	Unsecured Wi-Fi attack	Access confidential or personal data, such as banking or credit card information
	Phishing attack	Password-stealing and malware installation
	Poor authentication mechanism	Failure of software
	Malware attacks, i.e., Android OS	Increasing malware infection, e.g., Trojan, Rootkit, and Bot Process
Physical	Biometric authentication issues	Unauthorized device through physical access by using simple materials in direct and indirect way
	Physical attack	Failure of smartphone. Loss of asset
	Power and internet malfunction	Lack of controlling the smartphone
Network	Denial of service (DoS) and distributed denial of service (DDoS)	Stop smartphone system and communication
	Man-in-the-middle (MITM) attack	Unauthorized entity to smartphone system
	Sniffing and spoofing attack	Intercept user confidentiality and privacy. Insecure communications

Data theft or information stealing is the most possible attack in the first asset category. In discussions of data theft, the words "data breach" [49] and "data leak" might be used interchangeably. They differ, yet, in that a data leak happens when private information is unintentionally made public, whether online or by way of misplaced gadgets or hard drives. This makes it easy for malicious actors to obtain unauthorized access to private user information. A data breach, on the other hand, refers to deliberate cyberattacks rather than unintentional leakage [48, 50].

The *Application* category features four major areas of concern: unsecured Wi-Fi attacks, phishing attacks, poor authentication mechanisms, and malware attacks. Wi-Fi can be misused to monitor individuals using their devices, compromise passwords through phishing attacks, and disclose details about a person's employment or travels. Phishing attacks, on the other hand, can destroy systems, steal money and intellectual property, or install malicious software (like ransomware). Further complicating the issue, malware comes in a variety of forms, including spyware, ransomware, viruses, and adware, along with any other malicious software that can infiltrate a smartphone system.

This is followed by the *Physical* category, which consists of three concerns: biometric authentication issues, physical attacks (lost, theft, and damage), and power and internet malfunctions. One of the drawbacks of biometric systems is that, unlike passwords or ID tokens, they cannot be changed or revoked. This makes it exceedingly difficult, if not impossible, to replace a person's fingerprint or other physiological biometric if that trait has been compromised. In addition, the theft of a device allows an attacker to obtain physical access to a system or device.

The potential threats in the *Network* category include denial of service (DoS) attacks, distributed denial of service (DDoS) attacks, man-in-the-middle (MITM) attacks, and network sniffing and spoofing attacks. In a DDoS attack, a botnet of interconnected online devices is used to flood a target website with fake traffic. In contrast, a MITM attack is a form of eavesdropping in which the attacker interjects the full discussion before taking control of it. When an attacker creates TCP/IP using another person's IP address, this is known as spoofing.

1.4.3 Malware-on-the-Go

Mobile malware has emerged as the digital world's fastest growing cybersecurity threat. By inserting malicious code into a smartphone's operating system, mobile malware engages in malicious conduct that targets mobile and smartphones without the knowledge of the user. There are numerous ways for mobile malware to spread itself and attack smartphone operating systems. Some malware can infect systems by being attached as macros to files or by bundling with other programs through a variety of channels, including mobile network services, Internet access, and Bluetooth connections. The majority of mobile malware aims at mobile pick pocketing, which can involve stealing money or other valuables—via SMS and MMS— or may include the capacity to charge premium bills via SMS or calls.

Let's look at the comparison of Malware types including Ransomware, Banking Malware, Adware, Scareware, Spyware, and SMS malware in Table 1.2.

Table 1.2 demonstrates the variety of families on the common mobile OSes. Each malware category has a significant impact and mitigation:

- **Ransomware.** Mobile ransomware is modeled after desktop ransomware, which restricts device usage and demands a ransom from the affected user to restore control of the device or personal data. The victims of ransomware, including businesses and individuals, suffer from financial loss, reputational harm, and emotional effects. Ransomware attacks against hospitals have even resulted in patient deaths.[12]
- **Mitigation:** It is crucial to emphasize that the impact of ransomware can be minimized by periodically backing up all critical data.
- **Banking malware.** Mobile banking malware is a type of malware that is used to obtain access to an individual user's online banking accounts by imitating the original banking apps or mobile web interface. By deactivating the two-factor authentication system, attackers can circumvent the standard perimeter security features, such as web proxies and email gateways, and steal account credentials.
- **Mitigation:** Utilizing all the bank's security features is the viable approach in mitigating this attack.
- **Adware.** Adware is the term for the promotional content (i.e., advertisements) that frequently reside inside even the most trustworthy apps. Mobile adware asks the victim to install another app in a manner like scareware to begin its attack. Mobile adware then decreases the browsing performance and covertly collects user data.
- **Mitigation:** The easiest way to prevent adware is by installing the pop-up blockers on any mobile browsers.
- **Scareware.** Mobile scareware masquerades as legitimate programs and falsely claims to detect a wide range of risks on the affected mobile device (i.e., battery issues, malware threats). Scareware, such as ransomware, manipulates people by using human emotions, but instead of receiving money as a ransom, cybercriminals using a scareware attack obtain money as a product payment for the harmful software.
- **Mitigation:** Pop-up blockers, firewalls, and URL filters should all be used. In addition, instead of just dismissing the pop-up message, it is important to fully close the browser in order to mitigate the scareware.
- **Spyware.** Mobile spyware poses as a legal app in order to steal information and monitor a victim's behavior without their knowledge or permission. The software is not considered spyware if the end user is aware that monitoring software has been installed. Once infected with spyware, a malicious actor (or spy) can

---

[12] https://www.washingtonpost.com/politics/2021/10/01/ransomware-attack-might-have-caused-another-death/

1.4 Cybersecurity Challenges and Emerging Trends in Smartphones

Table 1.2 Understanding malware-on-the-go

Malware types	OS	Family variant	Impact	Mitigation
Ransomware	Android	Charger, Jisut, Koler, LockerPin, Simplocker, Pletor, PornDroid, RansomBO, Svpeng, WannaLocker	Industries and individuals face financial loss, damaged reputations, and emotional effects. Even deaths occurred because of ransomware attacks	• Only download files from trusted sources • Train and educate employee on how to spot the signs of Malware infections • Use multifactor authentication • Patch software regularly
	iOS	YiSpecter		
Banking malware	Android	Bankbot, Binv, Sandroid, Wroba, Fakebank, SMSspy, Zertsecurity, Citmo, Spitmo, Zitmo	Bypasses common perimeter security mechanisms, such as web proxies and email gateways and steal account credentials by removing the two-factor authentication system	Use all the security features offered by your bank
	iOS	Wirelurker		
Adware	Android	Dowgin, Ewind, Feiwo, Gooligan, Kemoge, Mobidash, Selfmite, Shuanet, Youmi	Slows down the browsing speed and collect data from the user without showing its presence in the user's system	Install a pop-up blocker
	iOS	Muda adware [51]		
Scareware	Android	AndroidDefender, AndroidSpy.277, AV for Android, AVpass, FakeApp, FakeAV, Penetho, FakeTaoBao, FakeJobOffer, VirusShield	Poses as legitimate apps and falsely claims to detect a variety of threats on the affected mobile device (i.e., battery issues, malware threats)	Use pop-up blockers, firewalls, and URL filters. Close the browser rather than just the pop-up notification
	iOS	Wirelurker, Rootpipe		
Spyware	Android	PhoneSpy, Pegasus, Sscul, Hermit	Poses as legitimate apps and secretly collect information and monitor activity	Use trusted antivirus software with antispyware features
	iOS	Trapsms, SantaAPT, Xsser mRAT, FinSpy Mobile, Spy App, InnovaSPY, SpyKey, mSpy, iKeyMonitor, FlexiSpy, OwnSpy, MobileSpy		
SMS malware	Android	BeanBot, Biige, FakeInst, FakeMart, FakeNotify, Jifake, Mazarbot, Nandrobox, Plankton, SMSsniffer, Zsone	Uses the SMS service as its medium of operation to intercept SMS payloads for conducting attacks	Use hardware tokens or two-factor authentication (2FA) apps
	Windows	D Anti-Terrorist game, PDA Poker Art, and Codec pack for Windows Mobile 1.0		

listen in on conversations held near a compromised smartphone or access data saved on, or transmitted by, the device.
- **Mitigation:** The use of reliable antivirus software with antispyware can mitigate this spyware attack.
- **SMS malware.** SMS malware intercepts SMS payloads to launch attacks using the SMS service as its operational platform. There are two types of SMS malware: SMS Phishing and SMS Fraud. SMS Phishing, often known as SMiShing, is a type of phishing that traditionally uses the social engineering methodology to retrieve sensitive information from users. Social engineering refers to any strategy that causes users to make a security mistake, such as sharing information they should not share, downloading software they should not download, visiting websites they should not visit, or any mistake that compromises their personal or devices' security. This strategy may be implemented through an SMS, email, or a voice call apparently from a trusted sender. SMS fraud is the abuse of the mobile premium service, a billing service for phones (SMS premium-rate). Due to its convenience as a mobile payment method, this service is favored by many reliable service providers.

- **Mitigation:** Using physical tokens or apps that require two-factor authentication can help in preventing this type of attack.

1.5 Chapter Summary

This chapter describes the historical evolution of smartphones and describes some of the main security issues relating to modern smartphones and their associated applications (apps). It explains how to understand the various attack types and explains the consequences of the attacks. The most important smartphone information security concerns—including data leaks, phishing, malware, outdated smartphone operating systems, and improper application permissions—are also covered in this chapter.

> By reading this chapter, you can answer the following questions:
> How many epochs has the smartphone traversed?
> What is the most popular mobile OS?
> What ways can you mitigate the security threats on smartphones?
> What are some common potential smartphone attacks, and what effects do they have?
> What are some of the names of the different types of malware families?
> Which type of malware is the most sophisticated?

References

1. Kanjo, E., Bacon, J., Roberts, D., & Landshoff, P. (2009). MobSens: Making smartphones smarter. *IEEE Pervasive Computing, 8*(4), 50–57.
2. Trifan, A., Oliveira, M., & Oliveira, J. L. (2019). Passive sensing of health outcomes through smartphones: A systematic review of current solutions and possible limitations. *JMIR mHealth and uHealth, 7*(8), e12649.
3. Mobile Design and Development. (n.d.). *O'Reilly Online Learning.* Retrieved January 11, 2022, from https://www.oreilly.com/library/view/mobile-design-and/9780596806231/ch01.html
4. Acs, Z. J., Song, A. K., Szerb, L., Audretsch, D. B., & Komlosi, E. (2021). The evolution of the global digital platform economy: 1971–2021. *Small Business Economics, 57*, 1629–1659.
5. Shakya, R. K., Rana, K., Gaurav, A., et al. (2019). Stability analysis of epidemic modeling based on spatial correlation for wireless sensor networks. *Wireless Personal Communications, 108*, 1363–1377.
6. O'Loughlin, K., Neary, M., Adkins, E. C., & Schueller, S. M. (2019). Reviewing the data security and privacy policies of mobile apps for depression. *Internet Interventions*, 110–115.
7. Cleary, G. (2018). *Mobile privacy: What do your apps know about you?* [Online]. Accessed 2023, from https://symantec-enterprise-blogs.security.com/blogs/threat-intelligence/mobile-privacy-apps
8. Pham, L. (2021). *Mobile application: Definition, technology types and examples 2023.* Accessed 2023, from https://magenest.com/en/mobile-application/
9. Ahvanooey, M. T., Li, Q., Rabbani, M., & Rajput, A. R. (2017). A survey on smartphones security: Software vulnerabilities, malware, and attacks. *International Journal of Advanced Computer Science and Applications, 8*, 30–45.
10. Dogtiev, A. (2023). *App stores list.* Accessed 2023, from https://www.businessofapps.com/guide/app-stores-list/
11. Alsmadi, I. (2019). Cyber security management. In *The NICE cyber security framework* (pp. 243–251). Springer.
12. Brook, C. (2023). *What is data integrity? Definition, types and tips.* Accessed 2023, from https://www.digitalguardian.com/blog/what-data-integrity-data-protection-101
13. Tan, Y. S., Ko, R. K. L., & Holmes, G. (2013) Security and data accountability in distributed systems: A provenance survey. In *IEEE international conference on high-performance computing and communications & 2013 IEEE international conference on embedded and ubiquitous computing.*
14. Hande, S. A., & Mane, S. B. (2015). An analysis on data accountability and security in cloud. In *International Conference on Industrial Instrumentation and Control (ICIC)*, Pune.
15. CIPL and Hodges, C. (2021). *Organizational accountability in data protection enforcement*, [Online]. Accessed 2023, from https://www.informationpolicycentre.com/uploads/5/7/1/0/57104281/cipl_white_paper_on_organizational_accountability_in_data_protection_enforcement_-_how_regulators_consider_accountability_in_their_enforcement_decisions__6_oct_2021_.pdf
16. Mayernik, M. S. (2017). Open data: accountability and transparency. *Big Data and Society, 4*(2), 1–5.
17. Hoboken, J. V., & Fathaighb, R. O. (2021). Smartphone platforms as privacy regulators. *Computer Law and Security Review, 41*.
18. Ma, X., Du, Z., & Liu, J. (2018). Program power profiling based on phase behaviors. *Sustainable Computing: Informatics and Systems, 19*, 341–350.
19. Amplifiers, W. (2022). *Cellular vs. Wifi: How safe is cellular data?.* Accessed 2023, from https://www.wilsonamplifiers.com/blog/cellular-vs-wifi-how-safe-is-cellular-data/#

20. Firoozjaei, M. D., Lu, R., & Ghorbani, A. A. (2020). An evaluation framework for privacy-preserving solutions applicable for blockchain-based internet-of-things platforms. *Security and Privacy, 131*.
21. Khana, J., Abbas, H., & Al-Muhtadi, J. (2015). Survey on mobile user's data privacy threats and defense mechanisms. In *International workshop on cyber security and digital investigation (CSDI 2015)*.
22. Delgado-Santos, P., Stragapede, G., Tolosana, R., Guest, R., Deravi, F., & VeraRodriguez, R. (2022). A survey of privacy vulnerabilities of mobile devices sensors. *ACM Computing Surveys, 54*(11), 1–30.
23. Baumgärtner, L., Dmitrienko, A., Freisleben, B., Gruler, A., Höchst, J., Kühlberg, J., Mezini, M., Mitev, R., Miettinen, M., Muhamedagic, A., Nguyen, T. D., Penning, A., Pustelnik, D., Roos, F., Sadegi, A., Schwarz, M., & Uhl, C. (2020). Mind the GAP: Security & privacy risks of contact tracing apps. In *IEEE 19th international conference on trust, security, and privacy in computing and communications (TrustCom)*.
24. Ali, A., Somroo, N. A., Farooq, U., Asif, M., Akour, I., & Mansoor, W. (2022). Smartphone security hardening: Threats to organizational security and risk mitigation. In *2022 International conference on cyber resilience (ICCR)* (pp. 1–12). IEEE.
25. Desai, M., & Jaiswal, S. (2020). Importance of information security and strategies to prevent data breaches in mobile devices. In *Improving business performance through innovation in the digital economy* (pp. 215–225). IGI Global.
26. Adăscăliței, I. (2019). Smartphones and IoT security. *Informatica Economica, 23*(2), 63–75.
27. A. (2020, October 7). *Top 8 mobile device cyber threats you should know to protect your data! Stealthlabs*. Retrieved February 2, 2022, from https://www.stealthlabs.com/blog/top-8-mobile-cybersecurity-threats-you-should-know-to-protect-your-data/
28. Hartrell, G. D., Steeves, D. J., & Hudis, E. (2012). *Malicious code infection cause and effect analysis*. https://patentimages.storage.googleapis.com/28/2d/57/2ab93c1faaf698/US8117659.pdf. US Patent 8,117,659
29. Mobile Techniques, MITRE ATT&CK. (2023). https://attack.mitre.org/techniques/mobile/ [online].
30. Clipboard Data, MITRE ATT&CK. (2023). https://attack.mitre.org/techniques/T1414/ [online].
31. Xu, E. & Guo, G. (2019). *Mobile campaign 'Bouncing Golf' affects Middle East*, [online]. https://www.trendmicro.com/en_us/research/19/f/mobile-cyberespionage-campaign-bouncing-golf-affects-middle-east.html
32. Gevers, R., Barbatei, A. M., Tivadar, M., Balazs, B., Bleotu, R., Coblis, C. (2019). *Uprooting mandrake: The story of an advanced Android Spyware Framework that went undetected for 4 years*. Bitdefender, [online], https://www.bitdefender.com/files/News/CaseStudies/study/329/Bitdefender-PR-Whitepaper-Mandrake-creat4464-en-EN-interactive.pdf
33. Lookout. (2019). *Monokle- the mobile surveillance tooling of the special technology center*, [online], https://www.lookout.com/documents/threat-reports/lookout-discovers-monokle-threat-report.pdf
34. Threat Fabric. (2019). *Cerberus - A new banking Trojan from the underworld*, [online], https://www.threatfabric.com/blogs/cerberus-a-new-banking-trojan-from-the-underworld.html
35. Snow, J. (2016). *Triada: organized crime on Android*, [online], https://www.kaspersky.com/blog/triada-trojan/11481/
36. Iarchy, R., & Rynkowski, E. (2018). *GoldenCup: New cyber threat targeting world cup fans*, [online], https://symantec-enterprise-blogs.security.com/blogs/expert-perspectives/goldencup-new-cyber-threat-targeting-world-cup-fans
37. Firoozjaei, M. D., Mahmoudyar, N., Baseri, Y., & Ghorbani, A. A. (2022). An evaluation framework for industrial control system cyber incidents. *International Journal of Critical Infrastructure Protection, 36*, 100487.
38. MITRE ATT@CK, Native API. Access in 2023, from https://attack.mitre.org/techniques/T1575/

References

39. Case, A., Lassalle, D., Meltzer, M., Koessel, S., Adair, S., Lancaster, T. (2020). *Evil eye threat actor resurfaces with iOS exploit and updated implant*, [online], https://www.volexity.com/blog/2020/04/21/evil-eye-threat-actor-resurfaces-with-ios-exploit-and-updated-implant/
40. Lookout. (2018). *Stealth Mango & Tangelo*. Security Research Report, [online], https://info.lookout.com/rs/051-ESQ-475/images/lookout-stealth-mango-srr-us.pdf
41. Hossain, M., Rafi, S., & Hossain, S. (2020). An optimized decision tree based android malware detection approach using machine learning. In *Proceedings of the 7th international conference on networking, systems, and security* (pp. 115–125).
42. Lookout. (2020). *Mobile APT Surveillance Campaigns Targeting Uyghurs*, [online], https://www.lookout.com/documents/threat-reports/us/lookout-uyghur-malwaretr-us.pdf
43. Flossman, M. (2017). *FrozenCell: Multi-platform surveillance campaign against Palestinians*, [online], https://www.lookout.com/blog/frozencell-mobile-threat
44. MITRE ATT@CK. *Matrix for Enterprise*. Access in 2023, from https://attack.mitre.org/
45. Guardsquare. (2017). *New Android vulnerability allows attackers to modify apps without affecting their signatures*, [online], https://www.guardsquare.com/blog/new-android-vulnerability-allows-attackers-to-modify-apps-without-affecting-their-signatures-guardsquare
46. Possemato, A., Aonzo, S., Balzarotti, D., & Fratantonio, Y. (2021). Trust, but verify: A longitudinal analysis of Android OEM compliance and customization. In *2021 IEEE symposium on security and privacy (SP)* (pp. 87–102).
47. Xiao, H. Z., Dong, Q., & Jiang, X. (2014). *Oldboot: The first bootkit on Android*. Qihoo 360 Technology Co. Ltd.
48. Hazum, A., He, F., Marom, I., Melnykov, B., & Polkovnichenko, A. (2019). *Agent Smith: A new species of mobile malware*, [online], https://research.checkpoint.com/2019/agent-smith-a-new-species-of-mobile-malware/
49. Husainiamer, M. A., Saudi, M. M., Ahmad, A., & Syafiq, A. S. M. (2021). Mobile Malware Classification for iOS Inspired by Phylogenetics. *International Journal of Advanced Computer Science and Applications, 12*(8).
50. Kondiloglu, A., et al. (2017). Information security breaches and precautions on Industry 4.0. *Технологический аудит и резервы производства, 6.4*(38), 58–63.
51. Zheng, C., Xiao, C., & Xu, Z. (2016). *New Android Trojan "Xbot" Phishes Credit Cards and Bank Accounts, Encrypts Devices for Ransom*, Security Research Report, [online], https://unit42.paloaltonetworks.com/new-android-trojan-xbot-phishes-credit-cards-and-bank-accounts-encrypts-devices-for-ransom/

Chapter 2
Android Operating System

Since its initial debut on September 23, 2008, Android has proven to be the most popular mobile operating system in the world. Android maintains its position today, with a market share of 71.93%, as of January 2022. You might be surprised to learn how much Android has evolved over time. The majority of people associate Android with smartphones; however, while smartphones are by far the most common device, Android is also seen on a variety of other devices as well. Currently, Android powers 2.5 billion active devices ranging from 5G smartphones to tablets, as well as smartwatches, televisions (TV), and cars [1].

2.1 Learning Basics: Android History

Android OS is built on top of the Linux 2.6 kernel and other open-source initiatives such as the Open Handset Alliance (OHA) and the Android Open-Source Project (AOSP). It was originally created by Android Inc. before being purchased by Google in 2005. Android is primarily intended for mobile devices, such as smartphones and tablets, which are battery-powered devices with hardware such as GPS receivers, cameras, light and orientation sensors, Wi-Fi and UMTS (3G telephony) connectivity, and a touch screen. The five components that make up Android are the *Linux Kernel*, *Android Runtime*, *Libraries*, *Application Frameworks,* and *Applications* as illustrated in Fig. 2.1.

The Android layered infrastructure ensures that the lower layers deliver services to the upper ones, as well as allowing programmers to focus on their own layers. For instance, the top layer of Android, called *Applications*, contains both built-in and third-party applications. The *Application Framework* layer provides a set of services for applications, allowing programmers to create a variety of applications by using the framework's Application Programming Interface (API). The *Android Runtime* is made up of core *Libraries*, which offer the majority of Java core library

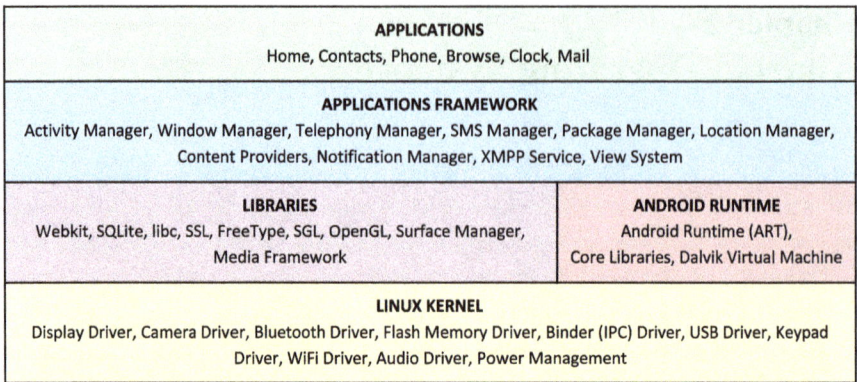

Fig. 2.1 Android system infrastructure (adapted from [2])

functionality, and a Dalvik Virtual Machine (DVM). Android may run numerous DVMs at the same time, each with its own application [3].

As you can see in Fig. 2.2, an Android application is packaged into an Android Package Kit (APK), which is a zip archive file type that contains a variety of files and folders, including the application code, resources, certificates, and manifest files. Android systems utilize a *.apk* file for the distribution and installation of mobile applications, similar to the way Windows (PC) systems use a *.exe* executable file for software installation. Google and the OHA have continued to develop Android since the initial release of the Android beta in 2007. Since then, it has received multiple modifications to its fundamental operating system.

Android had used a dessert-name theme starting from Android 1.5 until Android 9, with Android 10 utilizing a number-only scheme. The following is a list of all the major Android releases [4]:

- **Android 1.0:** The first commercial version was released on September 23, 2008, with basic features including a web browser, camera, Wi-Fi, Bluetooth, media player, Google synchronization, and other applications.
- **Android 1.1:** This version was released on February 9, 2009. Several bugs were fixed, the Android API was modified, and new functionality was added, including a longer in-call screen timeout, the ability to save attachments in messages, and the support for a marquee in system layouts.
- **Android 1.5 Cupcake:** The Android 1.5 upgrade was published on April 27, 2009. This was the first Android version to employ a codename based on a dessert item (Cupcake). Several new features and UI tweaks were introduced in this update such as the animated screen transitions and autorotation option.
- **Android 1.6 Donut:** Donut was released on September 15, 2009, and was based on the Linux kernel 2.6.29. Numerous new features were included in the upgrade such as speed improvements in searching and camera applications.
- **Android 2.0 Eclair:** Android 2.0 was released on October 27, 2009. The changes included improved hardware performance, a redesigned user interface, more screen sizes, and support for resolutions.

2.1 Learning Basics: Android History

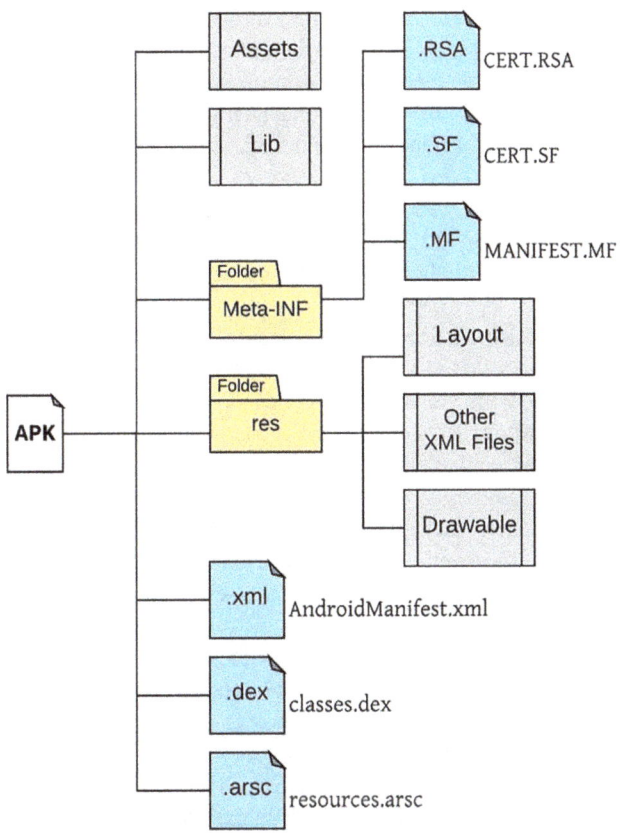

Fig. 2.2 The structure of an Android Package Kit (APK) [3]

- **Android 2.2 Froyo:** Froyo, short for frozen yogurt, was published on May 20, 2010, and was based on Linux kernel 2.6.32. Speed, memory, bug repairs, and performance optimizations were among the included enhancements.
- **Android 2.3 Gingerbread:** On December 6, 2010, Gingerbread was published. Support for Google Wallet, camera updates, enhanced battery, and user-redesigned interface were among the improvements.

- **Android 3.0 Honeycomb:** The Honeycomb (based on Linux kernel 2.6.36) was published on February 22, 2011, making it the first tablet-only Android upgrade. The update came with several features, including support for multicore processors, the ability to encrypt all user data, and HTTPS stack improvement with Server Name Indication (SNI).

- **Android 4.0 Ice Cream Sandwich:** On October 19, 2011, the Software Development Kit (SDK) for Android 4.0.1, which is based on Linux kernel 3.0.1, was published. This was the final version of Android that was certified to work with Adobe Systems' Flash player. Enhancements include improved stability, improved camera performance, smoother screen rotation, and improved phone number recognition.
- **Android 4.1 Jellybean:** Google announced Android 4.1 during the Google I/O conference on June 27, 2012. Jellybean was an incremental update based on Linux kernel 3.0.31, with the primary purpose of improving the functionality and performance of the user interface.
- **Android 4.4 KitKat:** KitKat was first released on Google's Nexus 5 on October 31, 2013, and was designed to run on a wider range of devices than previous Android versions, with 512 MB of RAM as a recommended minimum.
- **Android 5.0 Lollipop:** The Lollipop was launched under the codename "Android L." and was made available as an official over-the-air (OTA) update for select devices running Google's Android distributions on November 12, 2014. Lollipop came with a revamped user interface based on the "material design" responsive design language. Improvements to notifications, which could be accessed from the lock screen and presented as top-of-the-screen banners within apps, were among the new changes.
- **Android 6.0 Marshmallow:** On May 28, 2015, Google I/O unveiled Android 6.0 Marshmallow with the build number MPZ44Q for the Nexus 5 and Nexus 6 phones, Nexus 9 tablet, and Nexus Player set-top box under the codename "Android M."
- **Android 7.0 Nougat:** The operating system's seventh major update, Android Nougat, was released on March 9, 2016, which allowed supported devices to be upgraded to the Android Nougat beta via an over-the-air update.
- **Android 8.0 Oreo:** Android Oreo was the Android operating system's eighth major version. It was published on March 21, 2017, as a developer preview, called Android O, with factory images for Nexus and Pixel devices. On July 24, 2017, the final developer preview was released, followed by the stable version in August 2017. Android Oreo included a number of new features such as a native picture-in-picture mode, notification snoozing, and notification channels.
- **Android 9 Pie:** Android Pie was the ninth major update to the Android operating system. The first official release took place on August 6, 2018. The most significant change in Pie was the introduction of a hybrid gesture/button navigation system that replaced Android's standard Back, Home, and Overview keys with a large, multifunctional Home button and a small Back button that appeared alongside it, as needed.
- **Android 10:** On September 3, 2019, the stable version of Android 10 was launched. Most notably, the software introduced a completely redesigned interface for Android gestures, eliminating the tappable Back button entirely in favor of a swipe-based approach to system navigation.
- **Android 11:** Android 11 was released in September 2020. The most major changes in this edition concerned privacy: the update expanded on the extended permissions system introduced in Android 10 by allowing users to provide apps with permissions such as location access, camera access, and microphone access (just for a limited time and for a single use).

- **Android 12:** Android 12 is the twelfth major version of the Android operating system. Google released this final version of Android 12 in October 2021 and began sending it out to its Pixel devices (Pixel 6 and Pixel 6 Pro phones). In contrast to previous Android versions, the most significant advancements in Android 12 are mostly visual. Android 12 is the most significant redesign of Android's user interface since Android 5.0 (Lollipop).

2.2 Getting into Cybersecurity: Android Vulnerabilities and Risks

Your smartphone is unlikely to have the same level of protection as your work laptop or personal desktop PC. As a result, it is critical that you, the end user, do everything possible to defend yourself from cyber dangers. Getting into cybersecurity starts with becoming informed about the vulnerabilities and risks. Let's take a look at Table 2.1, which provides Android vulnerability statistics for all versions of Android with the following types of vulnerabilities [5–12]:

- **Denial of Service (DoS)**: DoS aims to bring down a computer system or network so that its intended users are unable to access it. DoS attacks achieve this by flooding the victim with an excessive amount of traffic, which leads to a crash.
- **Code Execution**: This flaw can be exploited remotely by an attacker to execute malicious code using commands. No matter where the device is physically located, remote code executions can happen.
- **Overflow**: This involves the execution of code as a result of a buffer overflow. A buffer, in this context, is a sequential area of memory that is set aside for the storage of anything from a character string to an array of numbers. Writing outside of the boundaries of a block of memory that has been allocated can corrupt data, cause a program to crash, or even be used to execute malicious code.
- **Memory Corruption**: This involves a vulnerability in computer systems that can happen when memory is changed without a clear assignment. Programming flaws cause the contents of a memory region to change, which allows attackers to execute malicious code.
- **SQL Injection (SQLi)**: SQL injection enables an attacker to alter the database queries that an application makes. SQL injections operate by taking advantage of holes in websites or computer programs, typically through data entry forms.
- **Cross-Site Scripting (XSS)**: XSS attacks take place when an attacker sends malicious code, typically in the form of a browser-side script, to a separate end user using an online application.
- **Directory Traversal**: Directory traversal enables an attacker to access any files on the server hosting an application. This could comprise critical operating system files, back-end system login information, and application code and data.

Table 2.1 Android vulnerabilities trends from 2011 to 2022 [5–12]

Year	Types of vulnerabilities										Total number of vulnerabilities
	DoS	CE	OF	MC	SQLI	XSS	DT	AB	IG	PE	
2011	✔	✔		✔		✔		✔	✔	✔	9
2012	✔	✔	✔						✔		7
2013		✔	✔	✔				✔	✔	✔	4
2014	✔	✔	✔		✔			✔	✔	✔	12
2015	✔	✔	✔	✔				✔	✔	✔	95
2016	✔	✔	✔	✔				✔	✔	✔	500
2017	✔	✔	✔	✔			✔	✔	✔	✔	840
2018	✔	✔	✔	✔	✔	✔	✔	✔	✔	✔	609
2019	✔	✔	✔	✔	✔			✔	✔	✔	491
2020	✔	✔	✔	✔	✔		✔	✔	✔	✔	859
2021	✔	✔	✔	✔	✔			✔	✔	✔	574
2022	✔	✔	✔					✔	✔	✔	67

DoS denial of service, *CE* code execution, *OF* overflow, *MC* memory corruption, *SQLI* SQL injection, *XSS* cross-site scripting, *DT* directory traversal, *AB* authentication bypass, *IG* information gain, *PE* privilege escalation

- **Authentication Bypass**: This attack takes advantage of weak authentication protocols to provide hackers with access to systems and data, including stealing legitimate session IDs or cookies to bypass the authentication system of a device.
- **Information Gain**: This involves permitting a local attacker, who has been authenticated, to obtain authentication details and gain unauthorized access to the system or database.
- **Privilege Escalation**: A privilege escalation attack aims to break into a system with privileged access without authorization. This happens when attackers take advantage of user error or design weaknesses in operating systems or web applications.

It is important to highlight that the majority of security concerns impacting Android systems are generated by applications rather than by the Android OS itself. This has resulted in a large number of mobile malware attacks. This is shown in the year 2020, which indicates the highest number of vulnerabilities, with a total of 859 vulnerability reports.

Figure 2.3 shows the evolution of Android security over a decade. *Google Play Protect*, the official Android market's security team, has been steadily improving its security measures in order to protect Android consumers. The *Google Play Protect* service is a robust threat detection service that monitors a device in real-time to protect its data and apps from malware.

Recently, Android began using both hardware and software security to build stronger defenses. The user is authorized with lock screen credentials, which begins the security process at the hardware level. Verified Boot helps to safeguard data in transit, and at rest, by ensuring that the system software has not been tampered with. Hardware-assisted encryption and key handling also help protect data in transit and at rest. For instance, Android 9 and higher support Android Keystore keys, which are

2.2 Getting into Cybersecurity: Android Vulnerabilities and Risks

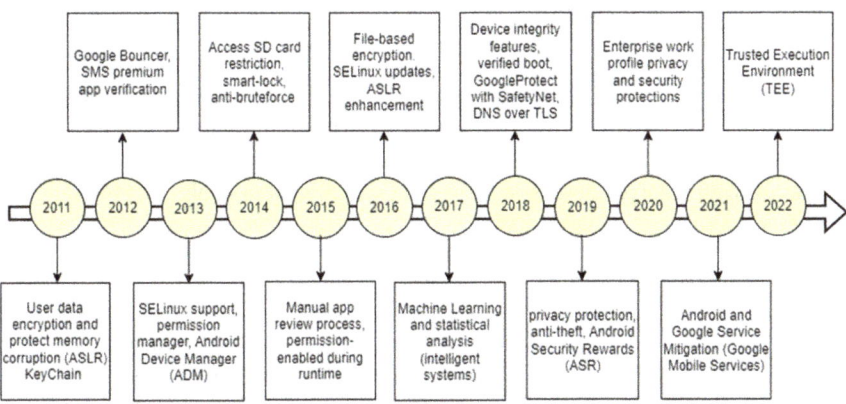

Fig. 2.3 Evolution of Android security features from 2011 to 2022 (Data are collected from Google blog at https://source.Android.com/security/)

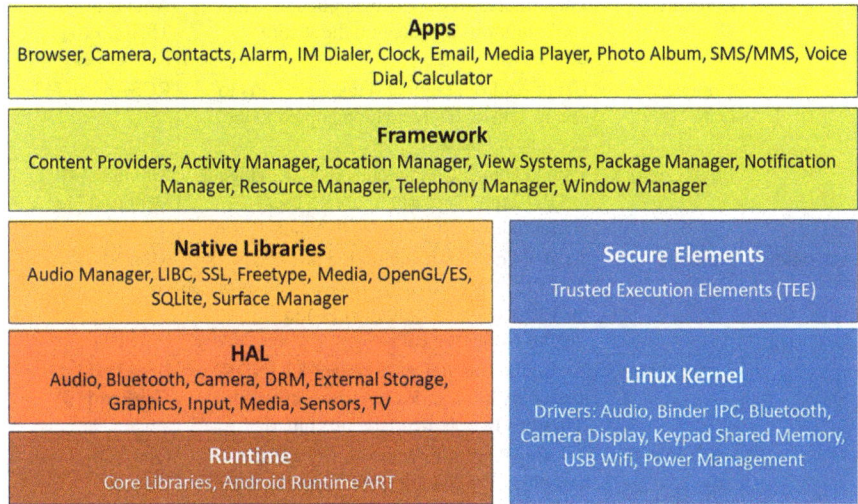

Fig. 2.4 The Android software stack: security by design [13]

kept and used in a physically separate CPU, designed specifically for high-security applications, such as an embedded secure element (SE), as depicted in Fig. 2.4.

A secure element service is a tamper-resistant execution environment on a chip that can securely execute applications and store data. Every Android phone has an SE on the universal integrated circuit card (UICC). The platform aims to provide users with the best experiences and the most up-to-date innovations while also keeping them safe by safeguarding their security and privacy.

2.3 Adversarial Techniques

Since Android is an open-source program and is highly customizable (unlike iOS), it is an easy target for cybercriminals who want to infiltrate a victim's mobile device using one or more of several adversarial techniques. Adversarial techniques used to compromise an Android device in these phases are listed in Table 2.2 [14].

Table 2.2 Adversarial techniques for Android OS [14]

Attack phase	Adversarial technique	Description	Sample
Propagation	Drive-By Compromise	Gaining access to a victim's Android device by visiting a compromised website	Stealth Mango [15]
	Network Connections Discovery	Utilizing Android APIs to discover nearby networks, e.g., Wi-Fi, Bluetooth, or cellular tower connections	XLoader [16]
	Software Discovery	Used to discover security applications and configurations installed on the victim's device	Anubis [17], TikTok Pro [1], Rotexy [2]
	Clipboard Data	Using Android clipboard manager APIs to obtain sensitive information copied to the device clipboard	RCSAndroid [3]
	Input Capture	Using capturing techniques (e.g., Keylogger or GUI Input Capture) to obtain user credentials or a user's sensitive data	Monokle [4], Dendroid [13]
Activation	Hooking	To avoid detection and hiding artifacts, hooking is used to modify return values or data structures of system APIs and the system's function calls	Monokle [4]
	Native API	Android's Native Development Kit (NDK) is abused by adversaries to write native functions to execute malicious binaries	Bread [18], HenBox [19]
	Input Injection	Abusing Android's accessibility APIs, a malicious application injects input into the user interface	Ginp [20], Gustuff [21]
	Software vulnerabilities Exploitation	Adversaries take advantage of a programming error in an application, service, within the OS software, or the kernel itself to execute adversary-controlled codes	Agent Smith [22], Gooligan [23]
	Impair Defenses	Adversaries maliciously modify components of a victim's mobile device in order to hinder or disable defensive mechanisms (e.g., anti-virus protection)	Zen [24], Dvmap [25]
	Hide Artifacts	Adversaries abuse Android's features and APIs to hide artifacts from the user to evade detection	Agent Smith [22], BusyGasper [26]

2.3 Adversarial Techniques

Table 2.2 (continued)

Attack phase	Adversarial technique	Description	Sample
Carrier	Adversary-in-the-Middle	Capturing or redirecting the network traffic of a mobile device by some solutions such as through a VPN agent, DNS redirection, or DNS poisoning	Monokle [4],
	Clipboard Data	Abusing Android's clipboard manager APIs to obtain users' sensitive information (e.g., password) copied to the device's clipboard	RCSAndroid [27, 28], GolfSpy [29]
	Application Layer Protocol	Blending malicious traffic (e.g., commands to smartphone or exfiltrate data) in the application layer protocol traffic, including web browsing, transferring files, electronic mail, or DNS traffic	DoubleAgent [30]
	Archive Collected Data	Obfuscating users' information collected from the smartphone by compressing or encrypting data	Golden Cup [31], GolfSpy [32], FrozenCell [33]
	Location Tracking	Tracking a mobile device's location through the use of standard Android APIs via malicious or exploited applications on the compromised device	Anubis [17], Android/Chuli.A [34]
Execution	Scheduled Task/Job	Abusing Android task scheduling functionality (e.g., WorkManager API) to facilitate initial or recurring execution of malicious code	GPlayed [35], Tiktok Pro [1]
	Interpreters abuse	Adversaries abuse command and script interpreters to execute malicious commands, scripts, or binaries	
	Endpoint Denial of Service	Abusing Android Device Administrator access to reset the device lock passcode, preventing the user from unlocking the device	Xbot [36], GPlayed [35], Exobot [37]
	Native API	Android's Native Development Kit (NDK) is abused by adversaries to write native functions to execute malicious binaries	Bread [18], HenBox [19]
	Data Manipulation	To manipulate external outcomes or hide the malicious activity, an attacker attempts to insert, delete, or alter the input or origin data	
Persistence	Compromise Application Executable	Injecting malicious code into an application that is installed on a user's Android mobile in order to carry out malicious tasks	Agent Smith [22]
	Hijack Execution Flow	Manipulating how Android OS locates the libraries or programs to be executed in order to execute the malicious payloads	
	Boot Initialization Scripts	Automatically executes malicious scripts at boot or logon initialization to establish persistence	OldBoot [38]

2.4 Dissecting Malware: Types of Android Malware

This section discusses Android malware and its types and includes how to tell whether your mobile device is infected, how a malware infection spreads, and what to do next.

Malware, short for "malicious software," is a software that is designed to harm a computer, a server, or a network. It is frequently found disguised in the form of useful applications, files, or media that do not appear to be overtly dangerous to casual observers. When the malware goes mobile, it can take on many different forms, including ransomware, spyware, adware, scareware, and banking malware. Any unwanted programs that are used to undertake unauthorized and dangerous activities on Android smartphones are referred to as **Android malware**.

Before the official release of the Android platform in 2008, there was hardly any mobile malware and the few investigations that were undertaken mainly concentrated on other prominent platforms. Symbian was the target of the first computer worm to attack mobile devices. Due to the quick growth of mobile platforms and the rise in mobile threats [39], the number of studies on mobile malware, notably Android malware, has been continuously rising. One of the first studies in this field, the work of Zhou et al., sought to provide researchers with an understanding of mobile malware through systematic characterization of the Android malware from a variety of angles. This work was subsequently followed by a number of more advanced detection methods that developed detection and mitigation techniques in a number of domains such as mobile botnet detection, detection of privacy violations, and detection of security policy violations.

Android malware gathers all valuable information and transmits it to remote servers by abusing apps, manipulating devices, and stealing sensitive data. The attackers get root privileges by exploiting other vulnerabilities such as app-level privilege escalation and kernel-level vulnerabilities. In addition, modern malware that exploits applications is built with mutation features such as polymorphism and metamorphism, resulting in a massive increase in the number of malware variants [18].

Despite the fact that various works have addressed these difficulties, the successful analysis of Android apps for malware detection is still a work in progress. To understand malware better, let's take a look at the Android taxonomy in Fig. 2.5. The taxonomy is developed based on the capabilities of the smartphone as well as the profit-making tactics of cybercriminals.

You might wonder how cybercriminals generate money from Android smartphones. The answer is, simply enough, by taking advantage of three different key elements of mobile devices:

- **Mobile services**: Mobile services, such as SMS, MMS, and Bluetooth, can be used as a platform for virus distribution.
- **Mobile usage**: Mobile usage can be controlled by preventing the user from accessing the device or some of the files stored on it; the user must pay a ransom

2.4 Dissecting Malware: Types of Android Malware

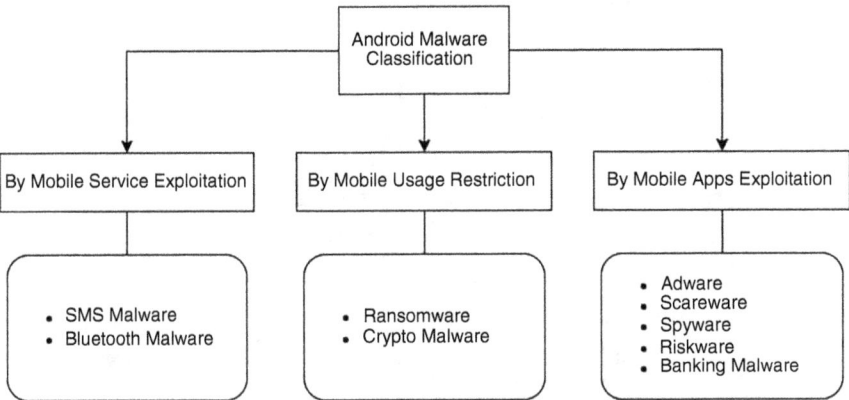

Fig. 2.5 Taxonomy of Android malware

to recover control of the device. Ransomware is a form of device exploitation that can be separated into two types: encryption-based and device-locking.

- **Mobile apps**: Unlike mobile service exploitation, mobile apps are frequently used as a vehicle for the rapid dissemination of malware as well as the exploitation of an app's features. Apps can be exploited in several ways (Fig. 2.6):

  - Matching the apps to the user's interests: when fraudsters take advantage of the app advertisement. The malware shows relevant advertising based on the user's interests. This is referred to as **adware**.
  - Encouraging the user to download infected apps: when cybercriminals profit from harmful apps by pressuring victims to download them or convincing them to pay a fee for a fictitious service. This method is known as **scareware**.
  - Posing a danger or harm to a user or an organization: Software that is vulnerable to cyber threats, both legitimate and illegitimate, and that may be secure for individual users but exposes businesses to data leakage, credential leakage, and information exfiltration that can be used to target specific employees in Advanced Persistent Threat assaults is referred to as **riskware**.
  - Posing as a legitimate app for learning yoga, seeing photos, watching TV, or other harmless activities, but in actuality, monitoring users' activities by endangering users' personal information, recording postings footage of the victim's screen to remote servers. This technique is known as **spyware**.
  - Imitating original apps: prominent apps such as Uber, WhatsApp, and Facebook, as well as financial apps, are frequently exploited in assaults. This type of malware is referred to as **banking malware**. Modern mobile banking malware, for example, not only steal financial information but can also intercept SMS conversations, and even lock and encrypt documents. Adware, spyware, and scareware can be used in conjunction with banking malware to obtain information about financial activities [3].

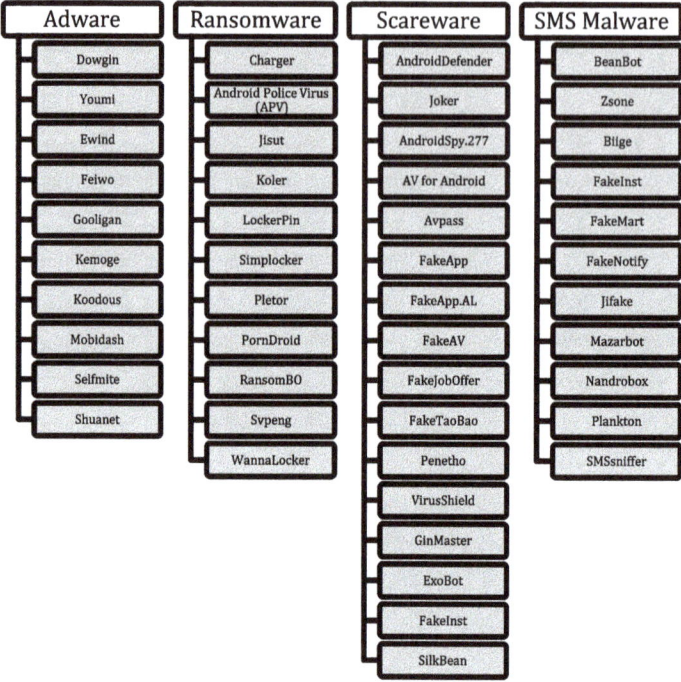

Fig. 2.6 Common Android malware categories and families [15, 16]

2.5 Mitigating Attacks: The Current Solutions

In the smartphone industry, Android OS-based devices hold a leading position. However, around 97% of mobile malware targets Android phones [40]. Android security relies on standard Unix and Java security paradigms that are the base of standard Linux distributions. Android processes are isolated from each other using different Unix process IDs, but applications can still communicate with each other by using so-called tents that transport exchanged information [39].

The amount of research on Android malware has been rapidly expanding with the development of mobile platforms and the increase in the number of mobile threats. The work by Zhou et al. [41] was one of the first in this field, and it attempted to give academics a better knowledge of mobile malware by systematically analyzing Android malware from numerous perspectives.

2.5 Mitigating Attacks: The Current Solutions

This research was rapidly followed by a number of more advanced detection methods aimed at establishing detection and mitigation techniques in a variety of areas, including mobile botnet detection, privacy violation detection, and security policy violation detection. Although different threats have been addressed and several detection techniques and tools have been developed, they all exhibit distinct limitations such that no single solution can claim to solve the Android malware problem [42]. Table 2.3 provides a summary of these studies.

Table 2.3 Comparison of popular existing Android malware study from 2011

Year	Authors	Focus area			Approach	Number of citation (Google Scholar)
		Analysis	Detection	Mitigation		
2011	Burguera et al. [43]		✔		Behavior-based malware detection	1387
2012	Zhou et al. [41]	✔			Behavioral analysis, characterization, and evolution	2647
2013	Peiravian et al. [44]		✔		Machine learning-based detection	385
2014	Yuan et al. [45]	✔	✔		Malware detection based on deep learning	434
2015	Tam et al. [46]		✔		Automatic reconstruction of malware behaviors	409
2016	Yuan et al. [5]	✔	✔		Analysis and detection by using deep learning	364
2017	Wei et al. [6]	✔			Deep analysis of Android malware	352
2018	Li et al. [7]		✔		Machine learning-based detection	397
2019	Chakkaravarthy et al. [8]	✔		✔	Survey on malware analysis and mitigation	76
2020	Alzaylaee et al. [9]	✔		✔	Detection by using a real device	148
2020	Rahali et al. [10]	✔	✔		Malware classification and characterization using deep image learning	39
2021	Imtiaz et al. [11]	✔	✔		Detection using deep learning ANN	38
2022	Amin et al. [12]		✔		Detection by using generative adversarial networks	21

2.6 Utilizing Android Services: The Trend Now

Mobile apps have become a necessary component of our daily lives. But what about businesses? In industrial zones such as factories, the power of mobile apps may boost production and quality. When we consider the various domains in which factories operate, we can see that there is a large opportunity for improvements that can be easily implemented by utilizing mobile phones. For instance, various Android tools are used for the Internet of Things (IoT) ecosystem, namely Node-RED,[1] OpenHub,[2] RIOT,[3] and Eclipse IoT[4] [41]. Android for Industry 4.0, Android Things, and Android Enterprise are among the current trends.

2.6.1 Android for Industry 4.0

Industry 4.0, in which devices are connected and communicate with one another via Android OS or another operating system, all while sending terabytes of data to the cloud, has arrived in recent years. They can even make judgments without involving humans, owing to next-generation technology such as virtual and augmented reality, artificial intelligence, neural networks, and machine learning.

In other words, those robotic arms are now able to communicate with one another and find out more effective ways to complete their tasks. It also extends beyond the assembly line. To ensure this type of connectivity, industrial apps will be critical. The reasons are obvious: applications are simple to use, cost-effective, and available in a user-friendly format everywhere. Any updates that are required can be easily obtained from the appropriate store. All these benefits lead us to believe that mobile apps have the potential to transform the way businesses operate. Let's look at some of those old, traditional methods and see how they might be supplanted by mobile apps.

2.6.2 Android Things

Android Things (previously known as Brillo) is a new IoT platform developed by Google that allows developers and end users to build devices using Android APIs and services. It includes the support library, which provides APIs to help anyone integrate with sensors, actuators, switches, and cloud services. Since the Android

---

[1] https://nodered.org/
[2] https://www.openhab.org/
[3] https://www.riot-os.org/
[4] https://iot.eclipse.org/

Things platform is an upgraded system of the basic Android, knowledge can be efficiently utilized to apply smart IoT-based applications [42].

Embedded board devices, e.g., energy meters, remote controls, and toys, run some kind of embedded OS that allows them to perform their tasks efficiently and reliably. Android Things tried to include the embedded system isolated devices into the Android ecosystem. Android Things is also an embedded OS (embedded Android) that is a special version of Android meant for devices that are not phones or tablets. Normally, in embedded systems, sensors gather information from the system, while actuators manipulate the variables in some way. The most obvious example of a smart device is a smart lock that has one sensor (a camera) and one actuator (a servo motor). This combination can be integrated with a biometric verification system that can operate remotely. Health-related machines are especially beneficial in the world of social distancing and pandemics. Other examples include smart lights, switches, surveillance drones, drone cluster light shows, and many more.

Although Android Things was designed to be secure, there were unsolved security concerns. The security and popularity issues caused Google to decide not to maintain the platform. Many device manufacturers failed to keep their hardware up-to-date with the latest updates for Android Things [41].

2.6.3 Android Enterprise

Google's Android Enterprise program enables the usage of Android devices and apps in the office. Developers can use the program's APIs and other tools to integrate Android support into their enterprise mobility management (EMM) systems. This site gives developers an overview of the program as well as the necessary background knowledge to begin developing an Android Enterprise solution. EMM console, Android Device Policy, and managed Google Play are the three components of an Android Enterprise system.

- **EMM console**: EMM solutions are usually implemented as an EMM console, which is a web application that allows IT administrators to manage their company, devices, and apps. You link your console with the APIs and UI components supplied by Android Enterprise to support these functions for Android.
- **Android Device Policy**: Android Device Policy must be installed on all Android devices managed through the EMM console during setup. Android Device Policy is an Android software that applies the management policies established in the EMM console to devices automatically.
- **Managed Google Play**: Managed Google Play for Business is an enterprise version of Google Play that simplifies app management for Android Enterprise solutions. It blends Google Play's familiar user interface and app store features with a collection of management tools intended exclusively for businesses. IT admins can use Managed Google Play to access services such as public app search, private app publishing, web app publishing, and app organization from their EMM

console. Managed Google Play is the user's enterprise app store on managed devices. Users can browse apps, read app details, and install them in a similar way to Google Play. Users can only install apps from controlled Google Play that their employer has approved for them, unlike the public version of Google Play.

2.7 Chapter Summary

This chapter describes Android vulnerability statistics for all versions of Android with the following types of vulnerabilities: Denial of Service, Code Execution, Overflow, Memory Corruption, SQL Injection, Cross-Site Scripting, Directory Traversal, Authentication Bypass, Information Gain, and Privilege Escalation.

The number of studies about Android malware has been rapidly expanding with the development of mobile platforms and the increase in the number of mobile threats. On the other hand, in industrial zones such as factories, the power of mobile apps may boost production and quality. Android for Industry 4.0 and Android Enterprise are among the trends now. There is a large opportunity for improvements that can be easily implemented by utilizing mobile phones in factories today.

> By reading this chapter, you can answer the following questions:
> Can you explain how an Android Package Kit is structured?
> How is the name of each Android version selected?
> What types of flaws exist in Android?
> How has Android security changed over the past 10 years?
> How can you determine whether your smartphone is infected?
> What are the current approaches to preventing Android attacks?

References

1. Desai, S. (2020). *TikTok Spyware, A detailed analysis of spyware masquerading as TikTok.* https://www.zscaler.com/blogs/security-research/tiktok-spyware
2. Shishkova, T., & Pikman, L. (2018). *The Rotexy mobile Trojan – banker and ransomware.* https://securelist.com/the-rotexy-mobile-trojan-banker-and-ransomware/88893/
3. Lashkari, A. H., Kadir, A. F., Taheri, L., & Ghorbani, A. A. (2018). Toward developing a systematic approach to generate benchmark Android malware datasets and classification. In *2018 International Carnahan Conference on Security Technology (ICCST)* (pp. 1–7).
4. Lookout, M. (2019). *The Mobile Surveillance Tooling of the Special Technology Center, Security research report.* https://www.lookout.com/documents/threat-reports/lookout-discovers-monokle-threat-report.pdf

5. Yuan, Z., Lu, Y., & Xue, Y. (2016). Droiddetector: Android malware characterization and detection using deep learning. *Tsinghua Science and Technology, 21*(1), 114–123.
6. Wei, F., Li, Y., Roy, S., Ou, X., & Zhou, W. (2017). Deep ground truth analysis of current Android malware. In *Detection of intrusions and malware, and vulnerability assessment: 14th international conference, DIMVA 2017*, Bonn, July 6–7, 2017, Proceedings 14 (pp. 252–276). Springer International Publishing.
7. Li, J., Sun, L., Yan, Q., Li, Z., Srisa-An, W., & Ye, H. (2018). Significant permission identification for machine-learning-based Android malware detection. *IEEE Transactions on Industrial Informatics, 14*(7), 3216–3225.
8. Chakkaravarthy, S. S., Sangeetha, D., & Vaidehi, V. (2019). A survey on malware analysis and mitigation techniques. *Computer Science Review, 32*, 1–23.
9. Alzaylaee, M. K., Yerima, S. Y., & Sezer, S. (2020). DL-Droid: Deep learning-based Android malware detection using real devices. *Computers and Security, 89*, 101663.
10. Rahali, A., Lashkari, A. H., Kaur, G., Taheri, L., Gagnon, F., & Massicotte, F. (2020). DIDroid: Android malware classification and characterization using deep image learning. In *2020 The 10th international conference on communication and network security* (pp. 70–82).
11. Imtiaz, S. I., ur Rehman, S., Javed, A. R., Jalil, Z., Liu, X., & Alnumay, W. S. (2021). DeepAMD: Detection and identification of Android malware using high-efficient Deep Artificial Neural Network. *Future Generation Computer Systems, 115*, 844–856.
12. Amin, M., Shah, B., Sharif, A., Ali, T., Kim, K. I., & Anwar, S. (2022). Android malware detection through generative adversarial networks. *Transactions on Emerging Telecommunications Technologies, 33*(2), e3675.
13. Lookout, *Dendroid malware taking over camera, record audio* (2014) https://www.lookout.com/blog/dendroid
14. MITRE Att@ck, Android Matrix. (2022). *Android Matrix.*
15. Lookout, Stealth Mango & Tangelo. (2018). *Selling your fruits to nation-state actors, Security research report.* Stealth Mango & Tangelo.
16. Hiroaki, H., Wu, L., Wu, L. (2019). *XLoader Disguises as Android Apps, Has FakeSpy Links.*
17. Feller, M. *Infostealer, Keylogger, and Ransomware in One: Anubis targets more than 250 android applications.* https://cofense.com/blog/infostealer-keylogger-ransomware-one-anubis-targets-250-android-applications/
18. Guertin, A., & Kotov, V. (2020). *PHA Family Highlights: Bread (and Friends), Android Security & Privacy Team, Google Security Blog.* https://security.googleblog.com/2020/01/pha-family-highlights-bread-and-friends.html
19. Hinchliffe, A., Harbison, M., Miller-Osborn, J., & Lancaster, T. (2018). *HenBox: The chickens come home to roost, Unit 42.* https://unit42.paloaltonetworks.com/unit42-henbox-chickens-come-home-roost/
20. Threat Fabric. (2019). *Ginp - A malware patchwork borrowing from Anubis.* https://www.threatfabric.com/blogs/ginp_a_malware_patchwork_borrowing_from_anubis.html
21. Pisarev, I. (2019). *Gustuff: Weapon of mass infection, Group-IB.* https://blog.group-ib.com/gustuff
22. Hazum, A., He, F., Marom, I., Melnykov, B., Polkovnichenko, A. (2019). *Agent Smith: A new species of mobile malware. Check Point Research.* https://research.checkpoint.com/2019/agent-smith-a-new-species-of-mobile-malware/
23. *More than 1 million Google accounts breached by Gooligan, Check Point Research Team* (2016). https://blog.checkpoint.com/research/1-million-google-accounts-breached-gooligan/
24. Siewierski, L., *PHA family highlights: Zen and its cousins*, Google Security Blog, https://security.googleblog.com/2019/01/pha-family-highlights-zen-and-its.html
25. Unuchek, R. (2017). *Dvmap: the first Android malware with code injection.* https://securelist.com/dvmap-the-first-android-malware-with-code-injection/78648/
26. Firsh, A. (2018). *BusyGasper – The unfriendly spy.* https://securelist.com/busygasper-the-unfriendly-spy/87627/

27. Trendmicro. (2015). *7 things you need to know about the Hacking Team's leaked mobile malware suite.* https://www.trendmicro.com/vinfo/us/security/news/mobile-safety/7-things-about-hacking-team-leaked-mobile-malware-suite
28. Fratantonio, Y., Bianchi, A., Robertson, W., Kirda, E., Kruegel, C., & Vigna, G. (2016). Triggerscope: Towards detecting logic bombs in android applications. In *2016 IEEE Symposium on Security and Privacy (SP)* (pp. 377–396).
29. Xu, E., & Guo, G. (2019). *Mobile Campaign 'Bouncing Golf' affects Middle East, Trendmicro.* https://www.trendmicro.com/en_us/research/19/f/mobile-cyberespionage-campaign-bouncing-golf-affects-middle-east.html
30. Lookout. (2020). *Mobile APT surveillance campaigns targeting Uyghurs.* https://www.lookout.com/documents/threat-reports/us/lookout-uyghur-malwaretr-us.pdf
31. Iarchy, R., & Rynkowski, E. (2018). *GoldenCup: New cyber threat targeting world cup fans.* https://symantec-enterprise-blogs.security.com/blogs/expert-perspectives/goldencup-new-cyber-threat-targeting-world-cup-fans
32. Xu, E. & Guo, G. (2019). *Mobile Campaign 'Bouncing Golf' Affects Middle East.* https://www.trendmicro.com/en_us/research/19/f/mobile-cyberespionage-campaign-bouncing-golf-affects-middle-east.html
33. Flossman, M. (2017). *FrozenCell: Multi-platform surveillance campaign against Palestinians.* https://www.lookout.com/blog/frozencell-mobile-threat
34. Baumgartner, K., & Maslennikov, D. (2013). *Android Trojan found in targeted attack.* https://securelist.com/android-trojan-found-in-targeted-attack-58/35552/
35. Ventura, V. (2018). *GPlayed Trojan - .Net playing with Google Market.* Talos. https://blog.talosintelligence.com/gplayedtrojan/
36. Zheng, C., Xiao, C., & Xu, Z. (2016). *New Android Trojan "Xbot" phishes credit cards and bank accounts, encrypts devices for ransom.* Security Research Report. https://unit42.paloaltonetworks.com/new-android-trojan-xbot-phishes-credit-cards-and-bank-accounts-encrypts-devices-for-ransom/
37. Threat Fabric. (2017). *Exobot (Marcher) - Android banking Trojan on the rise.* https://www.threatfabric.com/blogs/exobot_android_banking_trojan_on_the_rise.html
38. Vijay, A., Portillo-Dominguez, A. O., & Ayala-Rivera, V. (2022). Android-based smartphone malware exploit prevention using a machine learning-based runtime detection system. In *2022 10th International Conference in Software Engineering Research and Innovation (CONISOFT)*.
39. Kiss, N., Lalande, J.F., Leslous, M., & Tong, V.V.T., 2016. Kharon dataset: Android malware under a microscope. In *The LASER workshop: Learning from Authoritative Security Experiment Results (LASER 2016)* (pp. 1–12).
40. Tong, F., & Yan, Z. (2017). A hybrid approach of mobile malware detection in Android. *Journal of Parallel and Distributed Computing, 103*(2017), 22–31.
41. Zhou, Y., & Jiang, X. (2012, May). Dissecting android malware: Characterization and evolution. In *2012 IEEE symposium on security and privacy* (pp. 95–109). IEEE.
42. Razgallah, A., Khoury, R., Hallé, S., & Khanmohammadi, K. (2021). A survey of malware detection in Android apps: Recommendations and perspectives for future research. *Computer Science Review, 39*, 100358.
43. Burguera, I., Zurutuza, U., & Nadjm-Tehrani, S. (2011, October). Crowdroid: behavior-based malware detection system for Android. In *Proceedings of the 1st ACM workshop on security and privacy in smartphones and mobile devices* (pp. 15–26).
44. Peiravian, N., & Zhu, X. (2013, November). Machine learning for Android malware detection using permission and API calls. In *2013 IEEE 25th international conference on tools with artificial intelligence* (pp. 300–305). IEEE.
45. Faruki, P., Bharmal, A., Laxmi, V., Ganmoor, V., Gaur, M. S., Conti, M., & Rajarajan, M. (2014). Android security: A survey of issues, malware penetration, and defenses. *IEEE Communications Surveys and Tutorials, 17*(2), 998–1022.
46. Tam, K., Fattori, A., Khan, S., & Cavallaro, L. (2015, February). Copperdroid: Automatic reconstruction of Android malware behaviors. In *NDSS Symposium 2015* (pp. 1–15).

Chapter 3
iPhone Operating System (iOS)

Continuing our study on mobile OS, we now focus on Apple's iOS, the global leader in mobile systems. Even with a declining smartphone market in 2022, Apple's premium market share grew from 57% to 62% between Q1 2021 and Q1 2022, reinforcing its leadership [1].

3.1 Learning Basics: iOS History

iOS, formerly known as iPhone OS, is a proprietary, Unix-based mobile operating system developed by Apple Inc. It powers several Apple devices, including iPhones and iPod Touch, and forms the core for iPadOS, tvOS, and watchOS. Although some sections are open source, the OS is primarily proprietary and features a layered architecture (Fig. 3.1).

In a more sophisticated depiction, iOS is an intermediary between hardware and mobile applications. Instead of apps directly interacting with hardware, they connect via APIs, simplifying the creation of applications adaptable to different hardware specifications.

At the heart of iOS lies the Core Framework, forming the base for other technologies and offering low-level functionality. Briefly, the major components of iOS architecture [2] include the following:

- **Core OS:** Housing the kernel, file system, network setup, security, power management, and various device drivers. It also includes libSystem library supporting system-level APIs and POSIX/BSD 4.4/C99 API requirements.

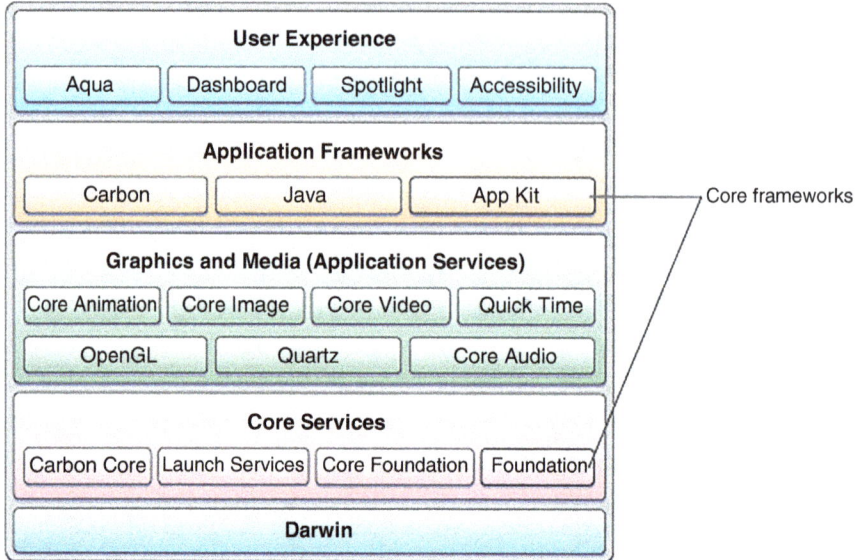

Fig. 3.1 iOS architecture [2]

- **Core Services:** Frameworks here provide core services based on device hardware characteristics such as GPS and accelerometer. They include System Configuration, Core Location, and Core Motion.
- **Media:** Relies on the Core Services layer to provide graphic and multimedia services to the Cocoa Touch layer, including video playback and Core Graphics.
- **Cocoa Touch:** This level offers frameworks directly supporting iOS-based applications, including the iAd and Game Kit.
- **UIKit:** Provides objects for user interface displays and dictates their behavior, including event handling and drawing.
- **Foundation:** Defines object management mechanisms and fundamental behavior and provides objects for fundamental data types, collections, and operating system services. The object-oriented variant of the Core Foundation is known as the Foundation.

3.2 Getting into Cybersecurity: iOS Vulnerabilities and Risks

While many perceive iOS devices as safer than others, their security is not foolproof. Aspiring cybersecurity professionals need to understand these vulnerabilities. Table 3.1 outlines key iOS security risks:

3.2 Getting into Cybersecurity: iOS Vulnerabilities and Risks

Table 3.1 iOS vulnerabilities trends from 2011 to 2022 (Data are collected from https://www.cvedetails.com/product/15556/Apple-Iphone-Os.html?vendor_id=49)

Year	Types of vulnerabilities										Total number of vulnerabilities
	DoS	CE	OF	MC	SQLI	XSS	DT	AB	IG	PE	
2011	✓	✓	✓	✓		✓		✓	✓	✓	83
2012	✓	✓	✓	✓		✓		✓	✓	✓	155
2013	✓	✓	✓	✓		✓	✓	✓	✓	✓	95
2014	✓	✓	✓	✓			✓	✓	✓	✓	120
2015	✓	✓	✓	✓				✓	✓	✓	386
2016	✓	✓	✓	✓		✓		✓	✓	✓	168
2017	✓	✓	✓	✓		✓		✓	✓	✓	388
2018	✓	✓	✓	✓		✓		✓	✓	✓	125
2019	✓	✓	✓	✓	✓	✓			✓	✓	356
2020	✓	✓	✓	✓		✓	✓	✓	✓	✓	305
2021	✓	✓	✓	✓		✓	✓	✓	✓	✓	365
2022	✓	✓	✓	✓				✓	✓	✓	43

DoS denial of service, *CE* code execution, *OF* overflow, *MC* memory corruption, *SQLI* SQL injection, *XSS* cross-site scripting, *DT* directory traversal, *AB* authentication bypass, *IG* information gain, *PE* privilege escalation

- **Denial of Service (DoS)**: A DoS attack intends to overwhelm a system with excessive traffic, causing it to crash and making it unavailable for intended users.
- **Code Execution**: An attacker can exploit a flaw from a distance to run malicious code. It is particularly concerning as it allows remote command execution, regardless of the device's location.
- **Overflow**: A specific form of code execution arising from a buffer overflow, which happens when writing exceeds the allocated memory block's boundaries. This can lead to data corruption, system crashes, or the execution of harmful code.
- **Memory Corruption**: This is a system vulnerability where memory changes occur without explicit assignment, usually due to programming errors. These changes can allow attackers to execute harmful code.
- **SQL Injection (SQLi)**: SQLi enables an attacker to manipulate database queries made by an application. These attacks often exploit gaps in websites or software programs, commonly through data entry forms.
- **Cross-Site Scripting (XSS)**: In XSS attacks, an attacker sends harmful code to another user through an online application. This typically takes the form of a browser-side script.
- **Directory Traversal**: This type of attack allows an attacker to access files on a server hosting an application, potentially compromising critical system files or application codes.

- **Authentication Bypass**: Here, attackers exploit weak authentication protocols to gain system access, sometimes using stolen session IDs or cookies to circumvent the device's authentication system.
- **Information Gain**: This attack allows an authenticated local attacker to acquire authentication details and gain unauthorized access to the system or database.
- **Privilege Escalation**: This attack focuses on unauthorized privileged access by exploiting user mistakes or system vulnerabilities. It typically involves attackers taking advantage of design flaws in operating systems or web apps.

3.3 Adversarial Techniques

While iOS boasts stronger defenses, it is not immune to various attack strategies. Compared to Android, fewer methods exist for iOS-focused malware. Table 3.2 outlines these techniques [18].

3.4 Dissecting Malware: Types of iOS Malware

While the security community has focused on malware on the Android platform, iOS applications have received significantly less attention owing to their closed-source nature. To address this gap, this section presents a summary of iOS malware and its types, including how to tell whether your mobile device is infected, how it spreads, and what can be done to prevent the infection.

Historically, malware on iOS has been limited to one of two scenarios: either you jailbroke your smartphone, hacking it to remove security limitations and installing malware as a result, or you were the target of a nation-state adversary. Table 3.3 groups iOS malware families according to their categorization, including spyware, adware, inception, click fraud, and stealer.

> **DID YOU KNOW?** YiSpecter is the first malware to exploit iOS' private APIs to attack non-jailbroken Apple iOS devices, which targets users who are based in China and Taiwan. The malware has the ability to: Install unwanted apps, force applications to show full-screen unwanted adverts, change Safari bookmarks and default search engines, and return user information to the server.

To understand iOS malware better, look at the iOS taxonomy in Fig. 3.2. The taxonomy is developed based on the types of tools found in the wild [12]. You might wonder how cybercriminals generate money from iOS smartphones despite the robustness of smartphone security. The answer is simply by taking advantage of four different tools:

3.4 Dissecting Malware: Types of iOS Malware

Table 3.2 Adversarial techniques for iOS

Attack phase	Adversarial technique	Description	Sample
Propagation	Drive-By Compromise	Gaining access to a victim's mobile device by visiting a compromised website (e.g., distributing malicious JavaScript and iframes to the victim's mobile)	INSOMNIA [3] and Stealth Mango [4]
	Supply Chain Compromise	Manipulating software dependencies and development tools (e.g., Xcode) or product delivery mechanisms to add malicious codes	XcodeGhost [5]
	Network Configuration Discovery	Collecting detailed network information about a connected iOS device, including the IMEI, IMSI, ICCID, serial number, and phone number through network configuration and settings	DualToy [6]
	Software Discovery	Discovering installed applications (e.g., the security apps) and their configurations on the victim's device by abusing private APIs in iOS devices	INSOMNIA [7] and YiSpecter [8]
	Clipboard Data	Accessing the *UIPasteboard.general.string* field in iOS (not for iOS 14 and later versions) in order to access the clipboard and obtain sensitive information that is copied there	XcodeGhost [5]
	Input Capture	Using capturing techniques (e.g., GUI Input Capture) to prompt a fake alert dialog in order to phish user credentials or sensitive data	TianySpy [2] and XcodeGhost [5]
Activation	Downloading dynamic codes	Evading static analysis checks, allowing adversaries to download and execute dynamic code through third-party libraries, such as JSPatch	Windshift [9], ZergHelper [10], and YiSpecter [8]
	Obfuscating Files or Information	Obfuscating payloads or files by compressing, archiving, or encrypting in order to avoid detection	WireLurker [1], TianySpy [2], and INSOMNIA [7]
	Geofencing	Abusing permissions to access location services in order to perform location-based actions, such as ceasing malicious behavior or showing region-specific advertisements. The iOS's built-in APIs can be used to set up and execute geofencing	eSurv [11] and Windshift [9]
	Software Vulnerabilities Exploitation	Taking advantage of a programming error in an application, service, or within the OS software or kernel itself to execute adversary-controlled codes	Pegasus for iOS [12] and INSOMNIA [7]
	Process Injection	Injecting malicious code into processes via *ptrace* (process trace) system calls in order to evade process-based defenses	INSOMNIA [7]

(continued)

Table 3.2 (continued)

Attack phase	Adversarial technique	Description	Sample
Carrier	Adversary-in-the-Middle	Capturing or redirecting a mobile device's network traffic by some solutions such as VPN agent, DNS redirection, or DNS poisoning	KeyRaider [13]
	Clipboard Data	Abusing the iOS's clipboard manager APIs (not for iOS 14 and later versions) in order to obtain the users' sensitive information (e.g., password), which has been copied to the device's clipboard	XcodeGhost [5]
	Application Layer Protocol	Blending malicious traffic (e.g., commands to smartphone or exfiltrate data) in the application layer protocol traffic, including web browsing, transferring files, electronic mail, or DNS traffic	Concipit1248 [14]
	Local System's file	Installing surveillanceware to be used for searching local system sources, e.g., file systems or local databases, to find sensitive data	Tangelo [15], Concipit1248 [14], and eSurv [11]
	Location Tracking	Tracking a mobile device's location through the use of standard Android APIs via malicious or exploited applications on the compromised device	eSurv [11], Pegasus for iOS [12], and Tangelo [15]
Execution	Scheduled Task/Job	Abusing iOS's task scheduling functionality to facilitate "init" the "ial" or recurring execution of malicious code or codes. For iOS devices, adversaries tend to abuse the NSBackgroundActivityScheduler API for low-priority operations that can run in the background [16]	
	Interpreters abuse	Abusing command and script interpreters to execute malicious commands, scripts, or binaries. In this way, it can be possible to steal information via injected malicious JavaScript	TianySpy [2]
	Subvert Trust Controls	Modifying code signing policies to enable execution of applications signed with unofficial or unknown keys. Policy modification can be done in a number of ways, including by Input Injection or malicious configuration profiles. For instance, in *YiSpecter*, fake Verisign and Symantec certificates were used to bypass iOS malware detection systems	YiSpecter [8], XLoader for iOS [17], Windshift [9]
	Exfiltration Over Alternative Protocol	Exfiltrating data over a different protocol than that of the existing command and control channel (e.g., FTP, SMTP, HTTP/S, DNS, SMB, or any other network protocol not being used)	TianySpy [2]

3.4 Dissecting Malware: Types of iOS Malware

Table 3.2 (continued)

Attack phase	Adversarial technique	Description	Sample
Persistence	Compromise Client Software Binary	Modifying system software binaries to establish persistent access to iOS devices. Executing these modified binaries enables the attacker to carry out malicious tasks	Pegasus for iOS [12]
	Hijack Execution Flow	Utilizing one or more physical connections (e.g., USB) to bypass the application store requirements and directly install malicious applications. For example, the *WireLurker* malware monitors iOS devices connected via USB to an infected OSX computer and installs downloaded third-party applications or automatically generates malicious applications onto the device	WireLurker [1], DualToy [6]
	Credentials from the Password Store	Collect keychain data from an iOS device in order to acquire credentials, as keychains are the built-in way for iOS to keep track of users' passwords and credentials	INSOMNIA [7]

Table 3.3 iOS malware types

Malware family	Category
TRacer, YiSpecter, wirelurker	Spyware
iOS adware Cydia, zerghelper	Adware
Inception APT, Inception Whatsapp	Inception
iPhone click fraud	Clickfraud
KeyRaider, AppBuyer, Xsser	Stealer

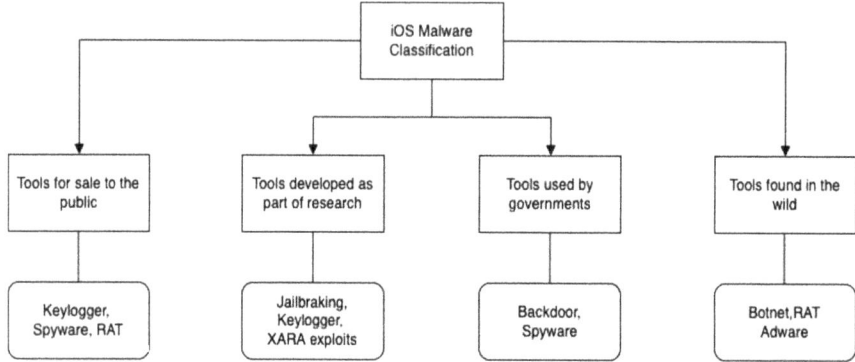

Fig. 3.2 iOS taxonomy based on the types of malware tools

1. **Tools for sale to the public:** This category aims to target individuals who use the three categories of tools, including keylogger, spyware, and RAT. The examples of available tools under this group are 1mole, copy9, copy10, FlexiSpy, iKeyMonitor keylogger, iKeyGuard Keylogger, InnovaSPY, Mobile Spy, MobiStealth, mSpy, ownSpy, Spy App, SpyKey, StealthGenie, and Trapsms.
2. **Tools developed as part of research:** The second category focuses on the research part as a proof of concept. This includes any malware tools developed by security researchers. For instance, one of the earliest tools, called iSAM, was developed in June 2011. This tool checks for jailbroken devices that are running SSH and still using the default password. Other types of tools under this research category include Instastock, Mactans, Jekyll, XARA, and NeonEggShell.
3. **Tools used by governments:** This category used by governments (and similar entities) exploits the two categories of attacks, backdoor and spyware, to target individuals such as activists and politicians. The following are a few examples of spyware tools commonly used by governments: FinSpy mobile, DROUPOUTJEEP, Hacking Team Tools, Inception, XAgent, Pegasus, Cellebrite, and CIA Vault 7.
4. **Tools found in the wild:** The last category targets the public and utilizes common attacks, including botnet, RAT, and adware. Some other examples include iKEE and Duh, Find and Call, Packages by Nobitazzz, AdThief/Spad, Unfold, AppBuyer, WireLurker, Xsser mRAT, Lock Saver Free, Keyraider, Muda, Youmi, AceDeceiver, XcodeGhost, YiSpecter, and main repo RAT.

3.5 Mitigating iOS Attacks: The Current Solutions

With the expansion of mobile platforms and the rise in mobile threats, the research volume on iOS malware has constantly increased. One of the pioneering works in this area was that of Felt et al. [19], who endeavored to improve academic understanding of mobile malware through survey analysis of malware behavior such as iOS, Android, and Symbian. This was followed by the work of Spaulding et al. [20], with the analysis of Wi-Fi vulnerability on the iOS platform. Table 3.4 summarizes the popular existing iOS malware study from 2011.

Table 3.4 Comparison of popular existing iOS malware study from 2011

Year	Authors	Focus area(s)			Approach	Number of citations (Google Scholar)
		Analysis	Detection	Mitigation		
2011	Felt et al.	✔		✔	Survey on malware behavior (iOS, Android, Symbian)	1077
2012	Spaulding et al.	✔			Analysis of Wi-Fi vulnerability as a malware attack vector	18
2013	Lau et al.	✔		✔	Malware injection via malicious charger	51
2014	Reinfelder et al.	✔		✔	Comparison studies of security and privacy between Android and iOS	33
2015	Bucicoiu et al.		✔	✔	Development of application sandboxing for hardening service	18
2016	D'Orazio et al.	✔		✔	Data exfiltration from IoT device (case study)	129
2017	Cimitile et al.	✔	✔		Machine learning-based detection	29
2018	Bhatt et al.	✔			Network forensic of iOS social networking apps	6
2019	Kellner et al.	✔	✔		Jailbreak detection on banking apps	11
2020	Husainiamer et al.	✔			Behavioral-based malware classification	0
2021	Bhatt et al.		✔		Active learning-based privacy leaks threat detection	3
2022	Hong et al.	✔			Systematic analysis of gambling scams	0

3.6 Utilizing iOS Services: The Trend Now

Apple's knack for innovation and reliable technology has spurred emerging trends in iOS app development. As Apple devices advance, these trends become integral to the app development process. The six dominant iOS app development trends for 2023 include wearable technology, mobile wallets, AR/VR, voice assistants, and app security.

3.6.1 Wearable Technology

In the realm of iOS app development, wearable technology is racing ahead. As of 2020, Apple's international smartwatch sales were expected to surpass 7.6 million units [11]. As demand continues to grow rapidly, we anticipate that sales of Apple smartwatches will remain strong.

3.6.2 Mobile Wallets (Use of Apple Pay)

As more people use smartphones, online payment systems are on the rise. Apple Pay offers security, one-click payment options, and convenience. It facilitates transactions via existing NFC technology in devices, avoids storing card details, and speeds up payment to merchants. Consequently, this trend in iOS app development is seeing a surge in usage.

3.6.3 Augmented Reality (AR) and Virtual Reality (VR)

AR and VR present striking innovations in iOS app development. By bridging the physical and digital worlds, they offer new ways to perceive information. In 2023, we expect further integration of AR and VR into iOS apps, enhancing customer engagement and user experience.

3.6.4 Voice Assistants

Currently, Siri assists users with tasks such as product searches via voice and basic troubleshooting. Apple introduced Siri in 2007 and integrated it into iOS, iPadOS, and watchOS. With SiriKit, integrating Siri into iOS-based apps is straightforward. Companies can also integrate Siri into their digital services to help customers with voice searches or basic issues.

3.6.5 iOS App Security

Apps on Apple devices are appreciated for their reliability and security. Given that iOS has stronger security measures than Android OS, these apps benefit from reliable security protections. In addition, this stringent policy enables safe storage of personal data. With concerns over cybercrime and malware attacks, app security is a key responsibility when developing iOS apps.

3.6.6 iOS HomeKit

HomeKit,[1] also known as Apple Home, is Apple's smart home platform, which is designed to let users control various internet-connected home devices ranging from thermostats and plugs to window blinds, light bulbs, and more using apps on their Apple devices, e.g., iPhone, iPad, Mac, or simple Siri voice commands. This framework links smart home products together and adds new capabilities to devices such as lights, locks, cameras, thermostats, plugs, and more [21].

> **DID YOU KNOW?** In their 2018 fiscal year, Apple sold more than 200 million iPhones. Apple releases a new iPhone model every September, and fans are excited to get the latest model. Customers are allowed to take advantage of iPhone trade-in schemes in order to upgrade more quickly.

3.7 Chapter Summary

This chapter provides an overview of iOS vulnerability statistics for all iOS versions such as Information Gain, Privilege Escalation, Denial of Service, Code Execution, Overflow, Memory Corruption, SQL Injection, Cross-Site Scripting, and Directory Traversal. With the growth of mobile platforms and the rise in mobile threats, the number of studies on the topic of iOS malware has been constantly increasing. As with Android systems, Apple is currently concerned about the inherent risks of mobile security threats. Although iOS is a target for mobile malware less frequently than Android, Apple users are known to take security breaches very seriously.

---

[1] https://developer.apple.com/documentation/homekit/

By reading this chapter, you can answer the following questions:
Can you explain the history of iOS?
Is it true that iOS is more secure than other available mobile OS options?
What changes have been made to iOS security over the last 10 years?
What are the current trends in iOS development?

References

1. Xiao, C. (2014). *WireLurker: A new era in OS X and iOS Malware*. https://unit42.paloaltonetworks.com/wirelurker-new-era-os-x-ios-malware/
2. Trend Micro. (2022). *TianySpy malware uses smishing disguised as message from Telco*. Trend Micro. https://www.trendmicro.com/en_us/research/22/a/tianyspy-malware-uses-smishing-disguised-as-message-from-telco.html
3. Case, A., Lassalle, D., Meltzer, M., Koessel, S., Adair, S., & Lancaster, T. (2020). *Evil eye threat actor resurfaces with iOS exploit and updated implant*. INSOMNIA. https://www.volexity.com/blog/2020/04/21/evil-eye-threat-actor-resurfaces-with-ios-exploit-and-updated-implant/
4. Lookout, Stealth Mango & Tangelo. (2018). *Selling your fruits to nation-state actors, Security research report*. Stealth Mango & Tangelo.
5. Xiao, C. (2015). *Novel malware XcodeGhost modifies Xcode, infects apple iOS apps and hits app store*. https://unit42.paloaltonetworks.com/novel-malware-xcodeghost-modifies-xcode-infects-apple-ios-apps-and-hits-app-store/
6. Xiao, C. (2016). *DualToy: New windows Trojan sideloads risky apps to android and iOS devices*. https://unit42.paloaltonetworks.com/dualtoy-new-windows-trojan-sideloads-risky-apps-to-android-and-ios-devices/
7. Case, A., Lassalle, D., Meltzer, M., Koessel, S., Adair, S., & Lancaster, T. (2020). *Evil eye threat actor resurfaces with iOS exploit and updated implant*. https://www.volexity.com/blog/2020/04/21/evil-eye-threat-actor-resurfaces-with-ios-exploit-and-updated-implant/
8. Xiao, C. (2015). *YiSpecter: First iOS malware that attacks non-jailbroken apple iOS devices by abusing private APIs*. https://unit42.paloaltonetworks.com/yispecter-first-ios-malware-attacks-non-jailbroken-ios-devices-by-abusing-private-apis/
9. Blackberry, B. (2020). *Hack-for-hire masters of phishing, fake news, and fake apps*. https://www.blackberry.com/us/en/pdfviewer?file=/content/dam/blackberry-com/asset/enterprise/pdf/direct/report-spark-bahamut.pdf
10. Xiao, C. (2016). *Pirated iOS app store's client successfully evaded apple iOS code review*. https://unit42.paloaltonetworks.com/pirated-ios-app-stores-client-successfully-evaded-apple-ios-code-review/
11. Lookout. (2019). *Phishing sites distributing IOS & Android surveillanceware*. Lookout Cloud & Endpoint Security. https://www.lookout.com/blog/esurv-research
12. Bazaliy, M., Flossman, M., Blaich, A., Hardy, S., Edwards, K., & Murray, M. (2016). *Technical analysis of Pegasus Spyware, an investigation into highly sophisticated Espionage Software*. Lookout. https://info.lookout.com/rs/051-ESQ-475/images/lookout-pegasus-technical-analysis.pdf
13. Xiao, C. (2015). *KeyRaider: iOS malware steals over 225,000 apple accounts to create free app Utopia*. https://unit42.paloaltonetworks.com/keyraider-ios-malware-steals-over-225000-apple-accounts-to-create-free-app-utopia/
14. Bao, T., Lu, J. (2020). *Coronavirus update app leads to project spy android and iOS spyware*. https://www.trendmicro.com/en_us/research/20/d/coronavirus-update-app-leads-to-project-spy-android-and-ios-spyware.html

References

15. Blaich, A., Kumar, A., Flossman, M., & Nickle, R. (2018). *Stealth Mango & Tangelo, Selling your fruits to nation-state actors, Lookout security research report.* https://info.lookout.com/rs/051-ESQ-475/images/lookout-stealth-mango-srr-us.pdf
16. Apple Developer, NSBackgroundActivityScheduler. https://developer.apple.com/documentation/foundation/nsbackgroundactivityscheduler
17. Hiroaki, H., Wu, L., & Wu, L. (2019). *XLoader disguises as android apps, has FakeSpy links.* https://www.trendmicro.com/en_us/research/19/d/new-version-of-xloader-that-disguises-as-android-apps-and-an-ios-profile-holds-new-links-to-fakespy.html
18. iOS Matrix, MITRE ATTACK Matrix, https://attack.mitre.org/matrices/mobile/ios/
19. Felt, A. P., et al. (2011). A survey of mobile malware in the wild. In *Proceedings of the 1st ACM workshop on security and privacy in smartphones and mobile devices.*
20. Spaulding, J., Krauss, A., & Srinivasan, A. (2012). Exploring an open WiFi detection vulnerability as a malware attack vector on iOS devices. *2012 7th International conference on malicious and unwanted software.* IEEE
21. Clover, J. (2022). *HomeKit: Everything you need to know.* https://www.macrumors.com/guide/homekit/

Chapter 4
Windows Operating System

The next technological evolution that we will cover in this book is Microsoft's Windows Phone, a mobile OS that has undergone significant changes over the years. The Windows Phone was first known as Windows Mobile in its early days until Microsoft recognized the need to adapt and innovate in response to the competitive landscape of the smartphone market. After the changes that were introduced by Apple (iOS) and Google (Android) in 2007, Microsoft decided to take a new direction and created Windows Phone as a response. This chapter delves into the history, evolution, and unique features of Microsoft's Windows Phone from its early beginnings as Windows Mobile to its latest updates and innovations as Windows Phone.

4.1 Learning Basics: Windows OS History

The development of Microsoft's mobile OS is a fascinating story of technological evolution and market competition. What factors led to the creation of Windows Phone, and how did it differ from its predecessor, Windows Mobile? How did Microsoft respond to the rise of Apple's iOS and Google's Android, and what impact did this have on the development of Windows Phone? These are just a few of the questions that arise when we consider the dynamic history of the mobile OS. Let's look at Fig. 4.1's timeline of significant Windows Mobile releases and events from 1999 to 2019 to get some answers to these queries.

Windows Mobile itself first launched in 1999 as the Pocket PC 2000, which operated on Pocket PC PDAs. However, Windows Mobile's origin story goes back to Windows CE, which was released in 1996 [1]. The evolution of Microsoft's mobile OS has gone through several name changes and updates over the years. In the early to mid-2000s, Windows Mobile 2003, Windows Mobile 2003 SE, and Windows Mobile 5.0 served as its foundation. In the late 2000s, Windows Mobile 6.0, 6.1, and 6.5 were introduced, and Windows Phone 7 and 7.5 (Mango) followed

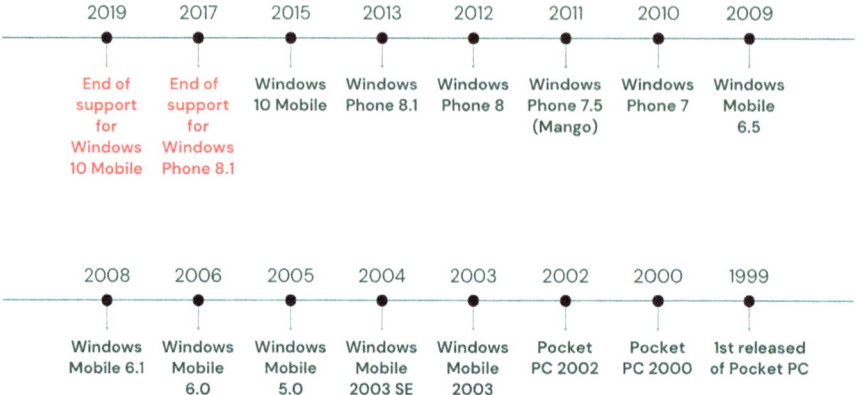

Fig. 4.1 Windows OS timeline from 1999 until 2019

in the early 2010s. The OS was renamed Windows 10 Mobile in 2015, with the release of Windows Phone 8 and 8.1 in 2012 and 2013, respectively. Microsoft's entry into the mobile phone industry, however, came to an end in 2017 when it stated that mainstream support for Windows Phone 8.1 and Windows 10 Mobile would expire in 2019 [2].

Throughout its development, the mobile OS underwent various improvements and changes to its user interface, its features, and its functionality in order to keep up with the rapidly evolving mobile landscape. The new interface, for example, was designed to be more modern, user-friendly, and optimized for touchscreens, as touchscreen technology was becoming more prevalent in smartphones. Additionally, Windows 10 Mobile was released with the intention of delivering a consistent experience across all Windows devices such as PCs, tablets, and smartphones. However, despite these efforts, Windows Mobile failed to gain a significant market share and eventually faced stiff competition from other mobile operating systems such as iOS and Android [3].

There are several factors that contributed to the decline and eventual termination of Windows Phone. Some of the key reasons are as follows:

1. **Poor sales performance**: The inability of Windows Phone to establish momentum in the market was one of the main causes of its eventual demise. IDC reported that Windows Phone only had 2.7% of the worldwide smartphone market in 2015, while Android had 81.5% and iOS had 15.9% [4].
2. **Lack of app support**: The lack of app support on Windows Phone is primarily a result of the platform's difficulty in attracting and retaining developers. Popular apps such as Snapchat and WhatsApp were not available or were supported poorly on Windows Phone, which was a major problem for users. As a result, users often turned to other platforms with more robust app ecosystems. In comparison to Android and iOS, there were a limited number of tools for the assurance of the extensibility, performance, scalability, and robustness of apps under

4.1 Learning Basics: Windows OS History

Windows OS [5]. The lack of such automatic assurance tools opens doors for adversaries for malicious purposes.

3. **Stiff competition**: Established smartphone platforms such as Android and iOS competed with Windows Phone. These platforms provided customers with a vast array of features and functionalities and had already cemented a solid position in the industry. This made it difficult for Windows Phone to differentiate itself and gain a competitive advantage.

Instead of focusing on building and maintaining its own mobile OS, the Microsoft company shifted its attention to creating applications and services for other mobile OS such as Android and iOS. Additionally, Microsoft is now focusing on offering enterprise mobility solutions through tools, such as Microsoft Endpoint Manager and Microsoft 365, which enable businesses to manage and secure their mobile devices, applications, and data. This move suggests that Microsoft is still invested in the mobile space but is choosing to approach it from a different angle that better aligns with its current strengths and market realities [6].

Although the Windows Phone architecture is similar to that of the Windows OS for desktop computers, there are some differences that arise as a result of the particular hardware and software requirements of the mobile platform. As shown in Fig. 4.2, the architecture of Windows Phone is a sophisticated system of components that work together to provide a seamless user experience. At the heart of this architecture are several crucial elements that are instrumental in ensuring the stability and performance of the platform. These elements include the Task Host, which manages the processing of background tasks, the Core Application, which provides a range of services to developers, and the Platform Services, which allow the applications to interact with the underlying hardware. Additionally, the Base OS Services provide a range of low-level services that are critical to the proper functioning of the platform. The following list provides an explanation of each component:

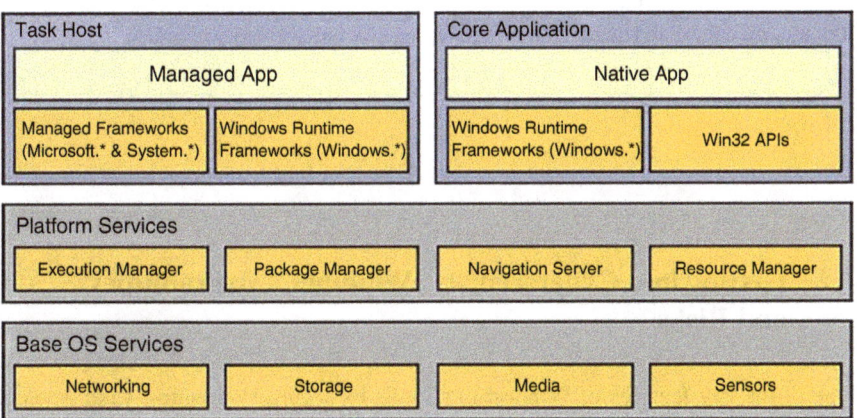

Fig. 4.2 Windows Phone architecture [c4-5]

1. **Task Host**: Although it is tailored for mobile devices, the Task Host in Windows Phone is similar to the one in desktop Windows. It controls the background operations and tasks, making sure that they function properly without affecting the system's overall performance.
2. **Core Application**: The Start screen, Phone, Messaging, and other basic applications that offer customers critical functionality are included in Windows Phone's basic applications. These applications are made to function perfectly with other elements of the OS and are tailored for the mobile platform.
3. **Platform Services**: Platform Services in Windows Phone provide a set of system-level services that support the OS and applications. These services include security services, networking services, device management services, and others. Like in desktop Windows, they are designed to be extensible so that developers can create custom services for their applications. The platform services have four subcomponents, namely Execution Manager, Package Manager, Navigation Server, and Resource Manager.
4. **Base OS Services**: In Windows Phone, Base OS Services include kernel-level services for managing system resources, drivers for hardware devices, and other low-level services that are essential to the operation of the system and ensure the OS functions effectively and are tailored for mobile devices. The Base OS Services consist of four subcomponents, namely Networking, Storage, Media, and Sensors.

The Windows Phone was mostly written using C++ and C#. Higher-level apps and user interfaces were created using C#, while lower-level system components were created using C++. For web-based applications, the platform also used other programming languages such as JavaScript and HTML5.

For instance, Windows Phone 7 is written in. NET managed code, which takes care of error-prone tasks. It supports two popular programming platforms, Silverlight and XNA, and development is done in Visual Studio. Programs are packaged into XAP files, which are Silverlight application packages. In conclusion, C++ and C# were combined to create Windows Phone, with support for JavaScript and HTML, five among other languages [7].

4.2 Getting into Cybersecurity: Windows Vulnerabilities and Risks

One of the key features of Windows Phone is its security measures. Like Apple's iOS, Windows Mobile OS takes a proactive approach to security by vetting and approving each piece of software that is uploaded to the Windows Store. This ensures that harmful programs cannot be downloaded onto the device, providing a safer and more secure user experience. In contrast to Android OS, Windows Mobile

4.2 Getting into Cybersecurity: Windows Vulnerabilities and Risks

Table 4.1 Windows Mobile OS vulnerabilities trends from 2006 to 2009 (Data are collected from https://www.cvedetails.com/product/9709/Microsoft-Windows-Mobile.html?vendor_id=26)

Year	Types of vulnerabilities									Total number of vulnerabilities
	DoS	CE	OF	MC	SQLI	XSS	DT	AB	IG	PE	
2006	✔	✔	✔								1
2007	✔		✔								4
2008	✔										1
2009		✔				✔					1

DoS denial of service, *CE* code execution, *OF* overflow, *MC* memory corruption, *SQLI* SQL injection, *XSS* cross-site scripting, *DT* directory traversal, *AB* authentication bypass, *IG* information gain, *PE* privilege escalation

does not require special antivirus or anti-malware software, further simplifying the user experience.

The Microsoft Windows OS was the safest mobile OS for enterprises (from 2006 until 2009) before it was discontinued in 2010. In contrast, Android continues to be the mobile device paradise for cybercriminals. Table 4.1 shows Windows OS vulnerabilities trends with the following categories of vulnerabilities:

- **Denial of Service (DoS)**: DoS aims to bring down a computer system or network so that its intended users are unable to access it. DoS attacks achieve this by providing the victim with an excessive amount of traffic that causes a crash.
- **Code Execution**: This flaw could be exploited remotely by an attacker to execute malicious code. An attacker can remotely execute commands. No matter where the device is physically located, remote code executions can happen.
- **Overflow**: This involves code execution as a result of a buffer overflow. A buffer, in this context, is a sequential area of memory set aside for the storage of anything from a character string to an array of numbers. Writing outside the boundaries of a block of memory that has been allocated can corrupt data, cause a program to crash, or even execute malicious code.
- **Memory Corruption**: This is a vulnerability in computer systems that can happen when memory is changed without a clear assignment. Programming flaws cause the contents of a memory region to change, allowing attackers to execute malicious code.
- **SQL Injection (SQLi)**: SQL injection enables an attacker to alter the database queries that an application makes. SQL injection operates by taking advantage of holes in websites or computer programs, typically through data entry forms.
- **Cross-Site Scripting (XSS)**: XSS attacks take place when an attacker sends malicious code, typically in the form of a browser-side script, to a separate end user using an online application.
- **Directory Traversal**: Directory traversal enables an attacker to access any files on the server hosting an application. This could comprise critical operating system files, back-end system login information, and application code and data.
- **Authentication Bypass**: This attack takes advantage of weak authentication protocols to provide hackers access to systems and data, including stealing legitimate session IDs or cookies to bypass the device's authentication system.

- **Information Gain**: This permits a local attacker who has been authenticated to obtain authentication details and gain unauthorized access to the system or database.
- **Privilege Escalation**: A privilege escalation attack aims to break into a system with privileged access without authorization. Attackers take advantage of user error or design weaknesses in operating systems or web applications.

4.3 Adversarial Techniques

In comparison to Android and Apple iOS, Windows Phone has become less popular among users. Few techniques have been detected to be used for adversarial purposes against the Windows Phone OSs. Since the same security mechanisms, which are used for Windows OSs (PC), are employed for protecting against emerging security threats for Windows Phone OSs [8], similar adversarial techniques are commonly used to compromise the Windows Phone OSs. There are so many third-party apps that are commonly available for Windows PCs and Phones, which make it possible for adversaries to employ Windows' malware to compromise smartphones. Any interaction between a Windows PC and a Windows Phone (e.g., software updating and file transferring) opens the door for an adversary. In Table 4.2, some of the adversarial techniques that are used to compromise a Windows Phone are listed.

Table 4.2 Adversarial techniques for Windows Phone [8]

Attack phase	Adversarial technique	Description	Sample malware
Propagation	Removable Media	Exploiting or copying malware to a Windows Phone connected to a Windows PC via USB	DualToy [9]
		Tracking a device's physical location through the use of standard OS APIs via malicious/spyware applications on the compromised device	
Activation	Privilege Escalation	Exploiting the software vulnerabilities, including a programming error in an application, service, OS's software, or kernel, to elevate privileges and execute an adversary-controlled code	FinFisher [10] and Wingbird [11]
Carrier	Web Protocols	Avoiding detection/network filtering by blending the malicious traffic with existing traffic (e.g., HTTP/S) or mobile messaging services (e.g., Google Cloud Messaging (GCM) or Firebase Cloud Messaging (FCM))	Dark Caracal (adds a registry key to the Windows folder or abuses Word documents macros) [12]
Persistence	Hijack Execution Flow	Abusing Windows' *KernelCallbackTable* is a process to hijack its execution flow in order to run the malicious payloads	FinFisher [10] and Wingbird [11]

4.4 Dissecting Malware: Types of Windows OS Malware

The emergence of mobile technology has brought with it a new set of security challenges, and Windows Phone OS is no exception. Malware designed for Windows phones can cause significant harm to users and their data. To understand the different types of Windows Phone malware, a taxonomy can be established based on the method of attack as shown in Fig. 4.3.

This taxonomy categorizes Windows Phone malware into four categories: Trojanized Gaming Applications, Code Execution, Man-in-the-middle Attacks, and Cross-Platform Viruses. In this taxonomy, each category represents a distinct method that malware authors can use to compromise Windows Phone OS devices. By understanding the methods of attack, users and security professionals can take steps to protect their devices against these types of malware.

- **Trojanized Gaming Applications:** Trojanized gaming applications are a common type of malware that masquerades as a legal game or gaming application in order to lure victims, particularly gamers. According to a report by NortonLifeLock, in 2020, the number of malware detections for gaming-related threats increased by 340% [13]. For instance, Windows-based smartphones have been infected by the malicious software "Dialer.BZ," which was built in 2015 and used to make illegal calls to premium-rate phone numbers, racking up expensive costs on the victim's phone bill. Microsoft stated that the attack affected users in countries such as Spain, Italy, Turkey, and India, and that they worked with local authorities to take down the infrastructure supporting the malware [14].
- **Code Execution:** Malware frequently uses code execution to compromise a target system and run malicious programs. This kind of attack can make use of vulnerabilities in the operating system or other software. The WannaCry ransomware outbreak, which hit thousands of computers around the world in 2017, is a well-known illustration of code execution [15]. Data from the victim was encrypted by the attacker, who took advantage of a flaw in the Microsoft Windows operating system and demanded a ransom to decrypt it. One incident

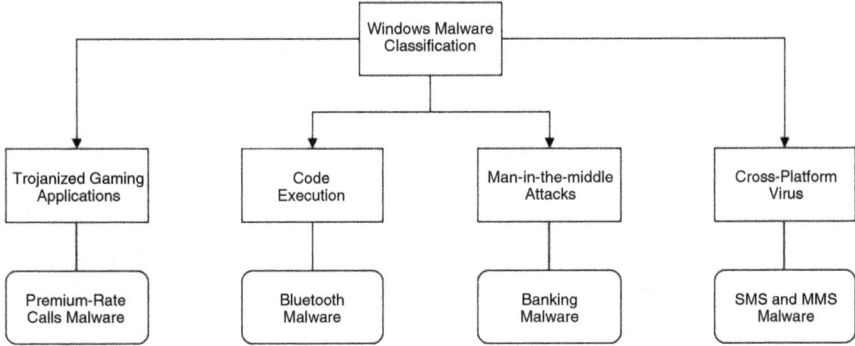

Fig. 4.3 Taxonomy of Windows Phone OS malware

that involved code execution through Bluetooth malware on Windows Phone was reported in 2014. The malware, called "Kemoge," was distributed through third-party app stores and disguised as popular apps such as Facebook, WhatsApp, and Twitter. Once installed on the victim's device, it would silently connect to a remote server and download additional malware [16].
- **Man-in-the-middle Attacks:** Man-in-the-middle (MITM) attacks are a type of cyberattack in which an attacker eavesdrops on and alters the communication between two parties in order to steal sensitive data or engage in malicious activity. According to a report by Akamai, MITM attacks increased by 54% in the first half of 2020 compared to the second half of 2019 [17]. MITM attacks with banking malware are a common threat to all mobile devices, including those running on Android and iOS. These attacks typically involve intercepting the communication between the user and their bank's server, allowing the attacker to steal the user's login credentials and other sensitive information. One such incident occurred in 2015 when a banking Trojan called "Acecard" was discovered to be targeting Windows Phone users in addition to Android users. The malware was designed to steal banking credentials and other sensitive information by overlaying fake login screens on top of legitimate banking apps. The Acecard Trojan was distributed through various channels, including spam emails and third-party app stores [18].
- **Cross-Platform Virus:** Cross-platform viruses are a type of malware that can infect several different operating systems or platforms, making them challenging to find and eliminate. The Mirai botnet, which targeted Internet of Things (IoT) gadgets, including routers, cameras, and DVRs, is one well-known example. The botnet was capable of infecting devices running various operating systems, including Windows, Linux, and Android [19]. There have been reports of SMS/MMS malware affecting Windows Phone in the past. One such incident occurred in 2003 when an SMS-based malware campaign called "SMSTrack" was discovered. The malware was spread through SMS messages that appeared to be legitimate tracking messages from popular delivery services but contained a malicious link that, when clicked, downloaded malware onto the device [20].

4.5 Mitigating Windows Attacks: The Current Solutions

Analyzing security threats (man-made or machine-made) is needed to identify and mitigate security attacks against mobile OSs. Modern information security solutions (e.g., machine learning-based approaches) rely on identifying anomalies that can identify false positive results, creating a sense of mistrust toward the system and thus requiring human effort to investigate cases. Effective artificial intelligence solutions can be used to improve situational awareness and implement effective protection measures [21]. A variety of artificial intelligence-based cybersecurity

4.5 Mitigating Windows Attacks: The Current Solutions

solutions are already introduced and reused for end devices (e.g., mobile devices, including Windows Phones) as follows:

- IBM MaaS360 Mobile Device Management (SaaS) [22]
- IBM Maas360 provides a cloud-based Unified Endpoint Management (UEM) solution designed to manage and secure a wide variety of endpoints and end users, including applications, content, and data. MaaS360 can be utilized on all major mobile computing platforms such as iOS, Android, and Windows Phones [21]. IBM MaaS360 Enterprise Mobility Management (EMM) tool is a mobile device management (MDM) tool introduced for Windows Phone devices. It has various features, including the management of applications, devices, browsers, and email. It provides some other unique solutions as well such as mobile expense management, secure sharing of documents, and mobile threats management [23].
- Deep Instinct [24]
- In this solution, deep learning algorithms are employed to identify structures used in malicious software. Deep Instinct can detect and prevent the execution of malicious software at all levels of the organization. To comprehensively analyze an attack, find how it has taken, and what kind of endeavors adversaries had, Deep Instinct has an On-Time Review and Remediate layer functionality, which provides visibility into the threats [21].
- SparkCognition DeepArmor
- SparkCognition, an artificial intelligence system, has launched its DeepArmor solution to cybersecurity, which uses machine learning to identify unknown data and detect cyber threats. DeepArmor aims at fixing the vulnerability of the end-points networks such as laptops, mobile devices, and sensors [25].
- IBM QRadar [26]
- QRadar Security Platform, a platform designed by IBM for information security analysis, reveals hidden threats and automates the authentication process of threats. This system provides automated threat investigations and uses Artificial Intelligence to detect high-level risks. QRadar implements local data mining of information security attacks by collecting relevant network data [21]. IBM QRadar Advisor with Watson provides automated research for threats and actors [27].

Based on available research, there has been a declining trend in the study of Windows Phone OS malware over the years. This is likely due to the decreasing market share and popularity of the platform, as well as the lack of major security incidents involving Windows Phone OS in recent years. As a result, there has been a shift toward studying other mobile platforms, such as Android and iOS, which have a larger user base and are more susceptible to security threats. Table 4.3 presents the comparison of existing studies with different focus areas such as analysis, detection, and mitigation.

Table 4.3 Comparison of existing Windows Phone malware studies from 2011 to 2022

Year	Authors	Focus area			Approach	Number of citation (Google Scholar) (The information is extracted in May 2023)
		Analysis	Detection	Mitigation		
2012	Ma et al.		✔		Anomaly-based Detection	191
2013	Zhang et al.	✔	✔		Static and Dynamic Analysis	102
2014	Yin et al.	✔			Dynamic Analysis	60
2015	Trinius et al.			✔	Firewall-based Mitigation	23
2016	Zaddach et al.		✔		Behavioral Analysis	68
2017	Alzubaidi et al.			✔	Permission-based Mitigation	8
2019	Wang et al.		✔		Static and Dynamic Analysis	5
2020	Ghaleb et al.			✔	Machine Learning	3
2021	Alhusain et al.		✔		Machine Learning	2

4.6 Utilizing Windows Mobile Services: The Trend Now

Unfortunately, Windows Phone has been discontinued by Microsoft since 2019, and there are no new trends in Windows Phone app development. Instead, Microsoft's current focus on mobile devices is on creating applications and services for other operating systems such as Android and iOS, as well as offering enterprise mobility solutions through tools such as Microsoft Endpoint Manager and Microsoft 365. The following are some of the trends in Microsoft Mobile Services:

- Developing cross-platform mobile apps using Xamarin, a tool for building native Android, iOS, and Windows apps with a shared codebase and user interface. For example, the mobile banking app of ANZ Bank uses Xamarin to provide a consistent user experience across different platforms.
- Leveraging Microsoft Endpoint Manager and Intune, cloud-based services for managing mobile devices and applications in the enterprise. For instance, Liberty Mutual Insurance Company uses Intune to deploy and manage apps on its employees' Windows phones and tablets.
- Building and deploying custom line-of-business apps on Windows phones and tablets, using tools such as Power Apps, Power Automate, and Power BI. These

apps can integrate with other Microsoft services such as Dynamics 365 and SharePoint. For instance, the sales team of Vodafone Netherlands uses Power Apps to access and update customer information on the go.
- Developing Internet of Things (IoT) solutions with Windows 10 IoT Core and Azure IoT services. For example, the HoloLens, a mixed-reality headset developed by Microsoft, uses Windows 10 IoT Core and Azure IoT to connect and control smart devices in a manufacturing plant.

4.7 Chapter Summary

Windows Phone OS is a mobile OS that was developed by Microsoft Corporation to compete with other mobile platforms such as iOS and Android. However, despite being praised for its unique and innovative design, it struggled to gain significant market share, and it eventually faded away. Throughout its development and existence, Windows Phone OS faced several security challenges, including malware attacks and vulnerabilities, which prompted researchers to conduct studies on various aspects of the platform's security. Some of the popular topics of research included malware detection, analysis, and mitigation.

Despite the decline of Windows Phone OS, the lessons learned from its development and security challenges remain relevant to the mobile industry. As mobile technology continues to evolve, it is important to stay up-to-date with the latest security trends and best practices to mitigate potential risks and threats. In conclusion, Windows Phone OS may no longer be a major player in the mobile market, but its history and security challenges provide valuable insights into the future of mobile security.

> By reading this chapter, you can answer the following questions:
> Can you explain the history of the Windows Phone OS?
> What factors led to the creation of Windows Phone?
> How did Windows Phone differ from its predecessor, Windows Mobile?
> How did Microsoft respond to the rise of Apple's iOS and Google's Android OS?
> What impact did this have on the development of Windows Phone?
> What caused Microsoft to discontinue the Windows Phone?

References

1. "Windows mobile OS: A brief history" by Sagar Khillar, *Interesting Engineering*, July 9, 2020. Available at: https://interestingengineering.com/windows-mobile-os-a-brief-history
2. "A brief history of Windows Mobile OS" by Russell Holly, *Android Central*, January 30, 2016. Available at: https://www.androidcentral.com/brief-history-windows-mobile-os
3. "A visual history of Windows Mobile" by Daniel Rubino, *Windows Central*, April 29, 2015. Available at: https://www.windowscentral.com/visual-history-windows-mobile
4. IDC. (2016). *Smartphone OS Market Share, 2015 Q4*. Retrieved from https://www.idc.com/promo/smartphone-market-share/os
5. Mohammad, D. R., Al-Momani, S., Tashtoush, Y. M., & Alsmirat, M. (2019). A comparative analysis of quality assurance automated testing tools for Windows mobile applications. In *2019 IEEE 9th Annual Computing and Communication Workshop and Conference (CCWC)* (pp. 0414–0419).
6. Jansen, W. (2021). *Microsoft 365 for business and enterprise*. Springer International Publishing.
7. Grønli, T. M., Hansen, J., Ghinea, G., & Younas, M. (2014, May). Mobile application platform heterogeneity: Android vs. Windows Phone vs. iOS vs. Firefox OS. In *2014 IEEE 28th international conference on advanced information networking and applications* (pp. 635–641). IEEE.
8. Ahvanooey, M. T., Li, Q., Rabbani, M., & Rajput, A. R. (2020). A *survey on smartphone security: Software vulnerabilities, malware, and attacks*. arXiv preprint arXiv:2001.09406
9. Xiao, C. (2016). *DualToy: New Windows Trojan sideloads risky apps to android and iOS devices*. https://unit42.paloaltonetworks.com/dualtoy-new-windows-trojan-sideloads-risky-apps-to-android-and-ios-devices/
10. Allievi, A., & Flori, E. (2018). *FinFisher exposed: A researcher's tale of defeating traps, tricks, and complex virtual machines*. https://www.microsoft.com/en-us/security/blog/2018/03/01/finfisher-exposed-a-researchers-tale-of-defeating-traps-tricks-and-complex-virtual-machines/
11. Microsoft. (2017). *Twin zero-day attacks: PROMETHIUM and NEODYMIUM target individuals in Europe*. https://www.microsoft.com/en-us/security/blog/2016/12/14/twin-zero-day-attacks-promethium-and-neodymium-target-individuals-in-europe/?source=mmpc
12. Lookout. (2018). *Dark Caracal: Cyber-espionage at a Global Scale*. https://info.lookout.com/rs/051-ESQ-475/images/Lookout_Dark-Caracal_srr_20180118_us_v.1.0.pdf
13. NortonLifeLock. (2021). *NortonLifeLock Cyber Safety Insights Report*. Retrieved from https://us.norton.com/internetsecurity-emerging-threats-cyber-safety-insights-report-volume-1-2021.pdf
14. Microsoft. (2015, July 23). *Dialer.BZ: Premium phone scam trojan on Windows phones*. Microsoft Security. https://www.microsoft.com/security/blog/2015/07/23/dialer-bz-premium-phone-scam-trojan-on-windows-phones/
15. BBC News. (2017). *WannaCry ransomware cyber-attacks slow but fears remain*. Retrieved from https://www.bbc.com/news/technology-39901382
16. Luo, Y., Zhu, H., Wang, Z., & Liu, P. (2016). Kemoge: Understanding mobile ad fraud in action. *Proceedings of the 33rd annual computer security applications conference (ACSAC)*.
17. Akamai. (2020). *State of the internet/security: Phishing for finance*. Retrieved from https://www.akamai.com/en/multimedia/documents/state-of-the-internet/state-of-the-internet-security-phishing-for-finance-report-2020.pdf
18. "Acecard banking trojan now targeting Windows Phone users," *SC Magazine*. (2015).
19. KrebsOnSecurity. (2016). *KrebsOnSecurity Hit with Record DDoS*. Retrieved from https://krebsonsecurity.com/2016/09/krebsonsecurity-hit-with-record-ddos/
20. Cuthbertson, A. (2016). *Windows Phone SMS malware threat discovered in Russia*. https://www.newsweek.com/windows-phone-sms-malware-threat-russia-449919
21. Vähäkainu, P., & Lehto, M. (2022). Use of artificial intelligence in a cybersecurity environment. In *Artificial intelligence and cybersecurity: Theory and applications* (pp. 3–27).
22. IBM, *IBM MaaS360 Mobile Device Management (SaaS)*, visited in 2023. https://www.ibm.com/docs/en/maas360

References

23. IBM, *Windows Phone 8 device MDM*, visited in 2023. https://www.ibm.com/docs/en/maas360?topic=windows-enrolling-your-phone-8-device-mdm
24. Deep instinct, https://www.deepinstinct.com/
25. Zhang, Y., Dai, Z., Zhang, L., Wang, Z., Chen, L., & Zhou, Y. (2020). Application of artificial intelligence in military: From projects view. In *2020 6th International Conference on Big Data and Information Analytics (BigDIA)* (pp. 113–116).
26. IBM Security QRadar Suite, IBM, https://www.ibm.com/qradar?utm_content=SRCWW&p1=Search&p4=43700074872917601&p5=e&gclid=CjwKCAjw1MajBhAcEiwAagW9MRLPpIdZxX5v4Cref8OJHY9QwQ35RD6wxdScMcFG4D4tMsWk7e5gpBoCNLIQAvD_BwE&gclsrc=aw.ds
27. Sadowski G, Kavanagh K, Bussa T (2020). *Critical capabilities for security information and event management*. Gartner Group Research Note.

Chapter 5
Other Operating Systems

You might be questioning why we would need an alternative operating system (OS) when most of the digital community is focusing on the trio of Windows, Android, and iOS. The answer to this question is straightforward: each OS has its own unique advantages and disadvantages, and, as a smartphone user, you have the flexibility to switch to an OS with the finest security, affordable prices, wider app selection, quicker software updates, and outstanding customer service.

In this chapter, we discuss the details of seven other popular operating systems: Symbian, Tizen OS, Sailfish OS, Ubuntu Touch, KaiOS, Sirin OS, and Harmony OS. These operating systems have all contributed to the diversification of the mobile OS landscape, offering unique features, security approaches, and user experiences. Let us delve first into the evolution of these mobile operating systems, as depicted in Fig. 5.1.

5.1 Learning Basics: The History

The development of each OS in this chapter (Symbian, Tizen, Sailfish, Ubuntu Touch, Kai, Sirin, and Harmony) is unique and caters to a range of devices, including smartphones, feature phones, and specialized devices. Each OS offers unique features, design principles, and user experiences, contributing to the diverse landscape of mobile computing. This section explores the key characteristics and highlights the architecture of these OS in the context of the evolving mobile technology ecosystem.

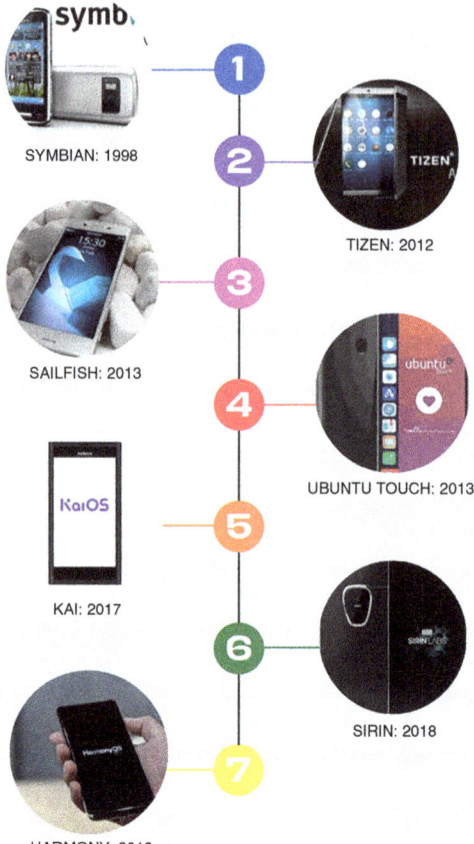

Fig. 5.1 Other alternatives mobile operating systems (from 2012 until 2019)

5.1.1 Symbian OS

In the 1990s, software company Psion was actively working on the development of innovative mobile operating systems. Their earlier products were 16-bit systems, but in 1994, they began working on a 32-bit version programmed in C++. This version was named EPOC32. In 1998, Psion rebranded Symbian Ltd in collaboration with popular mobile phone brands: Nokia, Ericsson, and Motorola. Symbian Ltd began upgrading EPOC32, and the new version was named Symbian OS.

The Symbian OS is most noteworthy for not having a known kernel compromise in its history, but it also implements an interesting security model [1]. In 2008, the Symbian OS was purchased by Nokia. Following the purchase, Nokia made the source code of Symbian OS publicly available under an open-source license. The use of the Symbian OS was discontinued in 2014 [2].

5.1 Learning Basics: The History

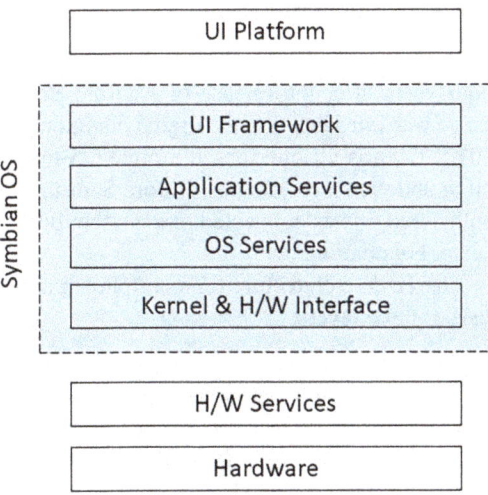

Fig. 5.2 Symbian OS architecture

Symbian OS is a 32-bit C++ − based, little-endian operating system. Further goals of the Symbian OS are for the CPU to have an integrated memory management unit (MMU) and a cache, which would allow it to operate in various privileged access modes and to handle interrupts and exceptions. The CPU, MMU, and cache, along with timers and hardware drivers, all reside on the system-on-chip [1].

The architecture of the Symbian OS is shown in Fig. 5.2 [3]. It separates the user interface (UI) from the engines and services, which allows manufacturers (e.g., Nokia and Sony Ericsson) to develop their own UIs for the phones.

The base component of Symbian OS is its kernel. Symbian uses a hard real-time, multithreaded microkernel architecture and has a request-and-callback approach to services. The kernel also performs preemptive multitasking. As shown in Fig. 5.2, Symbian OS consists of four layers, which are as follows [2]:

- **UI Framework layer**: The topmost layer of Symbian OS, the UI Framework layer, provides the frameworks and libraries for constructing a user interface.
- **Application Software layer**: The second layer of Symbian OS, the Application Software layer, provides services to applications and other higher level programs, independent of hardware, applications, or user interfaces. Services include specific application technology such as messaging and multimedia.
- **OS Service layer**: This layer, also known as the "Middleware layer," provides a framework, servers, and libraries.
- **Kernel Service layer**: The lowest layer of Symbian OS, the Kernel Services and Hardware Interface layer, contains the operating system kernel and the supporting components. These are responsible for abstracting the interfaces to the underlying hardware, including the logical and physical device drivers.

The Symbian OS is linked to mobile platforms (e.g., Nokia) through UI Platform and H/W Services layers as shown in Fig. 5.1.

5.1.2 Tizen OS History

Four years after the release of Android Version 1.0, in 2008, another OS called Tizen was introduced to the digital community. Created by Samsung Electronics in 2012, Tizen is a Linux-based mobile OS that emphasizes cross-platform compatibility and web technologies. Before Samsung combined its Bada operating system with Tizen, it was connected to a number of distributions and was supported by the Linux Foundation.

The Tizen architecture for smartphones and tablets is shown in Fig. 5.3 and consists of three layers:

1. **Application layer**: Tizen offers Web application support. Like native programs, Tizen Web applications take full advantage of the platform's capabilities.
2. **Core layer**: Tizen Core Service and Tizen API make up the Core layer. The Tizen Web API can be used to create Tizen Web applications. The Khronos WebGL, W3C (HTML5 and more), and newly defined device APIs are all included in the Tizen Web API.
3. **Kernel layer**: Device drivers and the Linux kernel are both included in the Kernel layer.

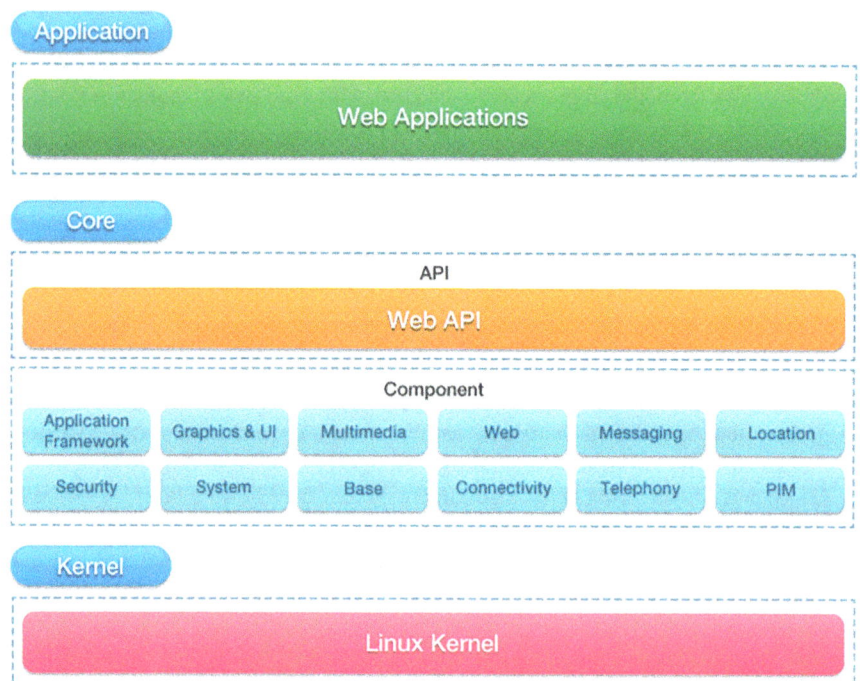

Fig. 5.3 Tizen architecture [4]

5.1.3 Sailfish OS History

Sailfish OS was developed by Sailfish Alliance, Mer, Jolla, and Sailfish community contributors in 2013. The OS is based on Nokia's MeeGo projects, which offered a gesture-based user interface (UI) known as "Sailfish UI" for smartphones, tablets, and wearables that enables simple navigation and interaction.

There are five essential components of Sailfish OS that contributes to its usability and user experience. These components include the following (Fig. 5.4):

1. **Mer Core**: The Mer Core, which acts as the operating system's fundamental layer, is based on Linux.
2. **Middleware**: Sailfish OS integrates middleware elements such as the Qt application framework, facilitating the execution of applications and ensuring uniform user interfaces.
3. **Silica**: The UI Framework, Silica, gives programmers access to user interface elements and controls.
4. **Alien Dalvik**: A proprietary technology developed by Myriad Group that enables compatibility with Android apps.
5. **Jolla Store**: The official app store, which facilitates app discovery and downloads, is also a feature of Sailfish OS.

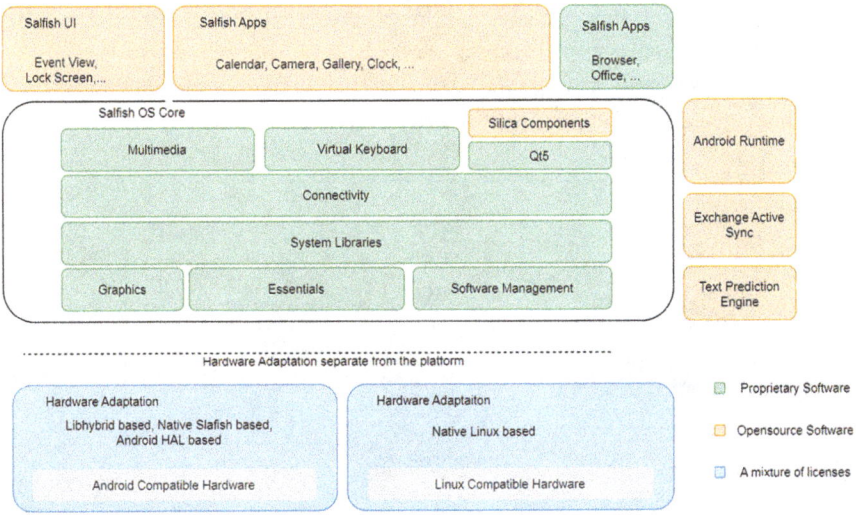

Fig. 5.4 Sailfish OS architecture [5]

5.1.4 Ubuntu Touch OS History

The Ubuntu Touch, also known as the Ubuntu Phone, is a mobile OS developed by Canonical Ltd, the company behind the Ubuntu Linux distribution. Canonical introduced Ubuntu Touch in 2013 with the goal of offering a unified experience across different forms of devices such as smartphones, tablets, and desktop computers. However, Canonical stopped work on Ubuntu Touch for smartphones in 2017.

Ubuntu Touch consists of various components, which are as follows (Fig. 5.5):

1. **Ubuntu Base**: The foundation of Ubuntu Touch provides the core operating system and essential components for running the system.
2. **Mir Display Server**: A display server that enables Ubuntu Touch to render graphical elements on supported devices and manage input and output operations.
3. **Unity 8**: The user interface shell for Ubuntu Touch, designed to provide a consistent and intuitive experience across different devices such as smartphones and tablets.
4. **Libertine**: A compatibility layer that allows Ubuntu Touch to run traditional Linux desktop applications, enabling users to install and use a wide range of software on their devices.

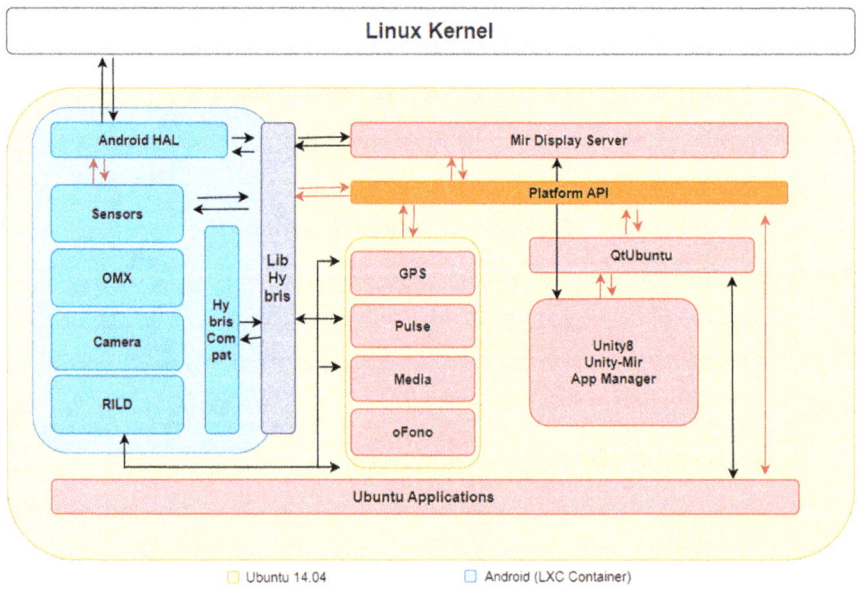

Fig. 5.5 Ubuntu Touch architecture [6]

5.1 Learning Basics: The History

5. **Ubuntu Store**: The official app store for Ubuntu Touch, offering a collection of applications specifically designed and optimized for the Ubuntu Touch platform.

5.1.5 KaiOS History

Various elements contribute to the KiOS architecture, which are as follows (Fig. 5.6):

1. **App Profile**: A component in KaiOS that defines the specifications and requirements for applications to run on the platform such as supported features, permissions, and compatibility guidelines.
2. **Web API**: This component provides a set of APIs that enable web-based applications in KaiOS to access device features, such as camera, contacts, and messaging, allowing developers to create interactive and feature-rich web apps.
3. **Core**: The central component of the KaiOS architecture manages system resources, handles system-level services, and coordinates communication between different components of the operating system.
4. **HAL (Hardware Abstraction Layer)**: This component acts as an intermediary layer between the core operating system and the hardware components of the device, providing a standardized interface for the operating system to interact with hardware functionalities such as sensors, displays, and input devices.

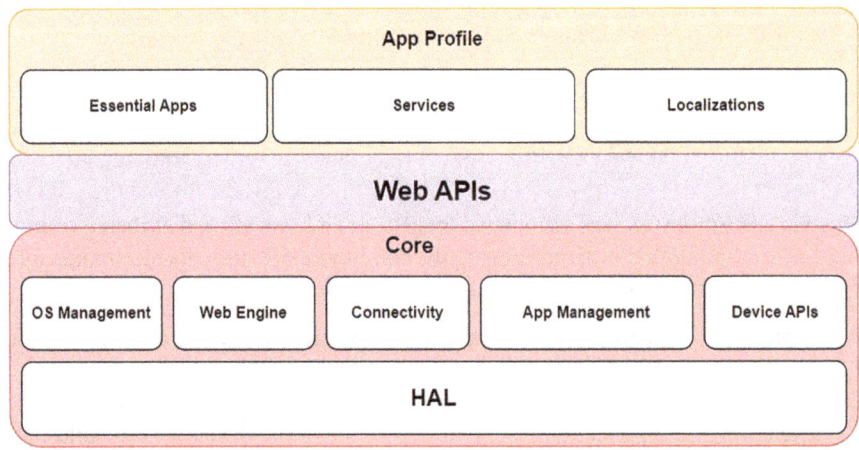

Fig. 5.6 KaiOS architecture [7]

5.1.6 Sirin OS History

Sirin OS, developed by Sirin Labs in 2018, is a mobile OS designed with a focus on security and privacy. Sirin OS is the only mobile OS that is secure enough to use and store cryptocurrency. The first blockchain smartphone is being developed by Sirin Labs, and all of their products are committed to utilizing their own blockchain.

Through SRN tokens, they encourage the usage of digital currencies and decentralization. Although specific information about the architecture of Sirin OS is not readily available, a general list of components is commonly found in mobile operating systems. These components may also be present in the architecture of Sirin OS:

1. **Kernel:** The core component of the Sirin architecture, responsible for managing hardware resources, providing low-level system services, and facilitating communication between software and hardware.
2. **Secure Enclave**: A specialized hardware component within Sirin devices that provides a secure and isolated environment for storing sensitive data, performing cryptographic operations, and enhancing overall device security.
3. **Security Framework**: A set of software modules and protocols implemented in Sirin to enforce security policies and handle authentication, encryption, access control, and other security-related functions.
4. **Behavioral-based IPS**: A security mechanism in Sirin that analyzes the behavior of applications and network traffic to detect and prevent potential threats such as malicious activities or intrusion attempts.
5. **Application layer**: The topmost layer of the Sirin architecture where user-facing applications and services are executed, providing the interface for users to interact with the device's features and functionalities.

5.1.7 HarmonyOS History

Huawei's HarmonyOS was announced in 2019 as an innovative, distributed operating OS for the Internet of Everything (IoE) era. Figure 5.7 presents the architecture of this OS as listed below:

1. **Application**: The top layer of HarmonyOS where user-facing applications are executed, providing functionality and interaction with the user.
2. **Framework**: A comprehensive set of libraries, APIs, and tools that facilitate the development of applications on HarmonyOS, enabling developers to build efficient and feature-rich software.
3. **System Service**: Core services provided by HarmonyOS that handle system-level operations such as device management, resource allocation, and inter-process communication, ensuring the smooth functioning of the operating system.
4. **Kernel**: The foundational layer of HarmonyOS that manages hardware resources, provides low-level system services, and facilitates communication between software and hardware components, ensuring proper device functionality.

5.1 Learning Basics: The History

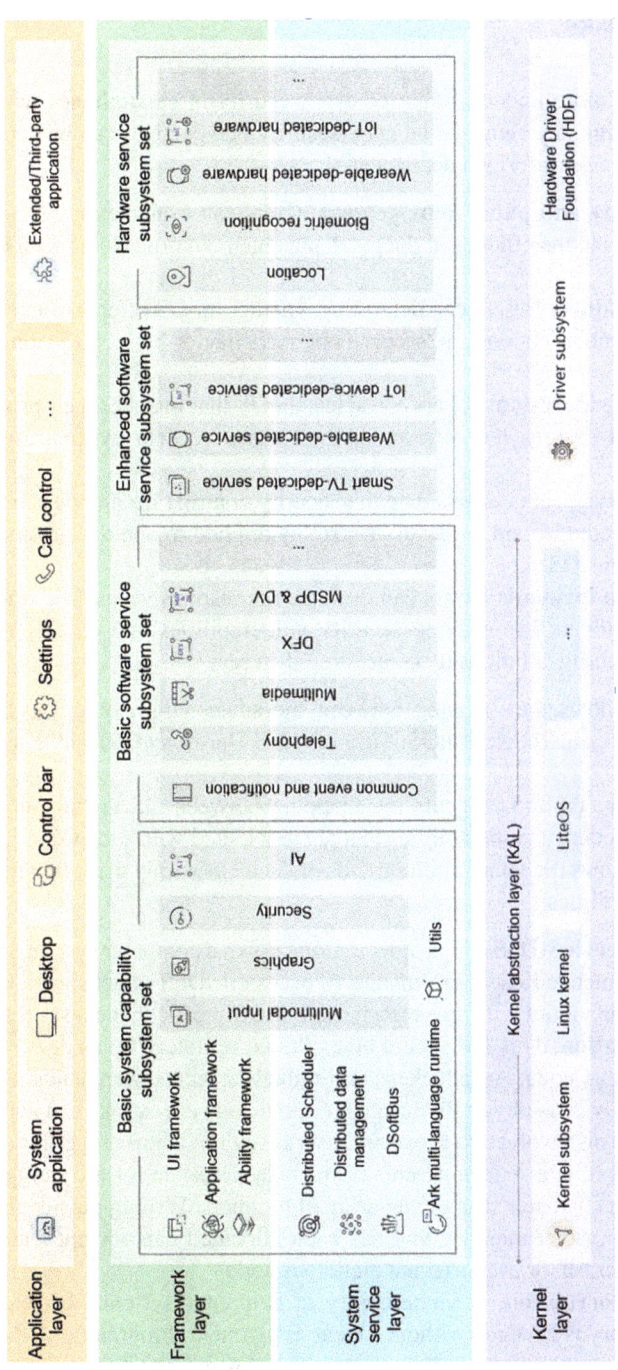

Fig. 5.7 HarmonyOS architecture [8]

5.2 Getting into Cybersecurity: The Vulnerabilities and Risks

In digging for more understanding of cybersecurity in smartphones with various OS, the following key elements on cybersecurity perspectives can be compared so that users can make informed decisions:

- **Vulnerability and patch management**: Users can evaluate the history of vulnerabilities and the efficacy of the patch management procedures used by the OS developers.
- **Security features**: Users can determine whether an OS offers stronger security measures such as encryption, secure boot, or behavior-based intrusion detection systems.
- **Privacy considerations**: Users can select an OS that matches their privacy preferences and ensures the security of their sensitive data by contrasting these features.
- **App security**: Users can evaluate the security controls for app distribution, app vetting procedures, and the level of scrutiny applied to app developers by comparing various OS.
- **Response to incidents**: Users can evaluate the responsiveness and track record of the developers in resolving security vulnerabilities and quickly delivering essential security solutions by contrasting different OS.

Table 5.1 shows the mapping between the seven OSs (Symbian, Tizen OS, Sailfish OS, Ubuntu Touch, KaiOS, Sirin OS, and Harmony OS) and their cybersecurity features.

Additionally, Table 5.2 details the mapping between OS architecture components and the security vulnerabilities that are associated with each OS.

Table 5.3 shows the vulnerability trends of each OS within the following categories of vulnerabilities:

- **Denial of Service (DoS)**: DoS aims to bring down a computer system or network so that its intended users are unable to access it. DoS attacks achieve this by providing the victim with an excessive amount of traffic that causes a crash.
- **Code Execution**: This flaw could be exploited remotely by an attacker to execute malicious code. An attacker can remotely execute commands. No matter where the device is physically located, remote code executions can happen.
- **Overflow**: This involves code execution as a result of a buffer overflow. A buffer, in this context, is a sequential area of memory set aside for the storage of anything from a character string to an array of numbers. Writing outside the boundaries of a block of memory that has been allocated can corrupt data, cause a program to crash, or even execute malicious code.
- **Memory Corruption**: A vulnerability in computer systems that can happen when memory is changed without a clear assignment. Programming flaws cause the contents of a memory region to change, allowing attackers to execute malicious code.

5.2 Getting into Cybersecurity: The Vulnerabilities and Risks 81

Table 5.1 Mapping between OS and cybersecurity features

Mobile OS	Vulnerability and patch management	Security features	Privacy considerations	App security	Response to incidents
Symbian	Regular security	Certificate management and cryptography	Granting access by the user	Integrity check (signing) and separation in kernel vs. user space	Responsive to security incidents
Tizen	Regular security updates	Secure boot, secure storage	User privacy controls	Strict app review process	Responsive to security incidents
Sailfish	Timely security patches	Encrypted storage, app sandboxing	User data control, app permissions	Security measures for app vetting	Responsive to security incidents
Ubuntu Touch	Prompt release of security patches	App confinement, encryption	User privacy controls, permissions	Security measures for app review	Responsive to security incidents
Kai	Regular security updates	App isolation, secure updates	User privacy controls	Strict app review process	Responsive to security incidents
Sirin	Information not widely available	Encryption, secure enclave	Emphasizes user privacy	Information not widely available	Information not widely available
Harmony	Regular security updates	Trusted execution environment	User privacy controls	Security measures for app review	Responsive to security incidents

- **SQL Injection (SQLi)**: SQL injection enables an attacker to alter the database queries that an application makes. SQL injection operates by taking advantage of holes in websites or computer, typically through data entry forms.
- **Cross-Site Scripting (XSS)**: XSS attacks take place when an attacker sends malicious code, typically in the form of a browser-side script, to a separate end user using an online application.
- **Directory Traversal**: Directory traversal enables an attacker to access any files on the server hosting an application. This could comprise critical operating system files, back-end system login information, and application code and data.
- **Authentication Bypass**: This attack takes advantage of weak authentication protocols to provide hackers access to systems and data, including stealing legitimate session IDs or cookies to bypass the device's authentication system.
- **Information Gain**: This permits a local attacker who has been authenticated to obtain authentication details and gain unauthorized access to the system or database.
- **Privilege Escalation**: A privilege escalation attack aims to break into a system with privileged access without authorization. Attackers take advantage of user error or design weaknesses in operating systems or web applications.

Table 5.2 Mapping between architecture components and their security vulnerabilities

OS (year)	Key features	Architecture components	Security vulnerabilities
Symbian (1998)	• C++ is the primary language for software development • Multitasking and preemptive kernel	Four layers: User Interface Framework, Application Software, OS Service layer, and Kernel Service layer	Lack of stability when presented with corrupted or nonstandard system files, vulnerable to overwrites system applications with corrupted ones, no concept of roles or users, and no access controls in the file systems [9–11]
Tizen (2012)	• Developed in HTML5 • Supports several Samsung wearables and smartphones.	Three components: Application, Core, Kernel	Lack of third-party app support, the potential for security vulnerabilities in the web runtime (insufficient isolation mechanisms, allowing applications to interfere with each other or access sensitive user data)
Sailfish (2013)	• Fully independent of Android. • Traces its roots from Nokia and MeeGo OS	Five components: Mer Core, Middleware, Silica, Alien Dalvik, Jolla Store	Limited app ecosystem, the potential for security vulnerabilities in third-party apps (may not receive regular security updates or patches)
Ubuntu Touch (2013)	• Connected to the Linux Open Store and independent from Google and Android services • Supports smartphones and tablets	Five components: Ubuntu Base, Mir Display Server, Unity 8, Libertine, Ubuntu Store	Limited app availability, the potential for security vulnerabilities in Libertine (privilege escalation and code injection)
Kai (2017)	• Linux-based • Support analog cell phones • Reduced energy use	Four components: App Profile, Web API, Core, Hardware Abstraction Layer (HAL)	Limited functionality compared to full-fledged operating systems. HAL vulnerabilities can include buffer overflow and input validation issues
Sirin (2018)	• Specifically designed to be sufficiently secure to hold cryptocurrencies • Good for crypto traders • Owns a tokenization system (SIRIN Token)	Five components: Kernel, Secure Enclave, Security Framework, Behavioral-based IPS, Application layer	Potential security vulnerabilities in Android subsystems, reliance on the Android app ecosystem
Harmony (2019)	• Huawei's Android challenger • Huge potential • Will primarily target the Chinese market	Four components: Application, Framework, System Service, Kernel	Potential vulnerabilities in the app ecosystem, lack of third-party app support

5.3 Dissecting Malware: The Type of Malware

Table 5.3 Mobile OS vulnerabilities trends

Mobile OS	Year	Types of vulnerabilities										Total number of vulnerabilities
		DoS	CE	OF	MC	SQLI	XSS	DT	AB	IG	PE	
Symbian	2009	✔	✔		✔				✔			5
Tizen	2021		✔									4
Sailfish	2018		✔									1
Ubuntu Touch	2016										✔	3
Kai	2020					✔						
Sirin												
Harmony (Hongmeng)	2021											266
	2022	✔	✔	✔		✔		✔	✔	✔	✔	

DoS denial of service, *CE* code execution, *OF* overflow, *MC* memory corruption, *SQLI* SQL injection, *XSS* cross-site scripting, *DT* directory traversal, *AB* authentication bypass, *IG* information gain, *PE* privilege escalation

5.3 Dissecting Malware: The Type of Malware

In contrast with the previous chapters, this section highlights the limitation of malware analysis for each OS, namely Symbian, Tizen OS, Sailfish OS, Ubuntu Touch, KaiOS, Sirin OS, and Harmony OS. It is important to note that the limitations mentioned in this section do not imply that these OS are inherently less secure or more prone to malware. Rather, it highlights the potential challenges faced in terms of dedicated resources, research focus, and response speed in malware analysis for niche OS that are less widely used.

1. **Symbian OS**: Due to its design and structure, Symbian OS was affected by a variety of mobile malware. Some of these well-known malwares were Cabir (a worm), Duts and Skuller (both viruses), Drever, and Locknut (both Trojans) [10].
2. **Tizen OS**: Malware analysis for Tizen OS may be limited due to its relatively smaller market share compared to other major OS. As a result, there may be fewer resources and dedicated research focused on analyzing and detecting Tizen-specific malware.
3. **Sailfish OS**: Similarly, Sailfish OS has a niche market presence, which can result in fewer security researchers actively analyzing and reporting on malware targeting this OS. This limited attention may lead to delayed detection and response to new or evolving malware threats.
4. **Ubuntu Touch**: Ubuntu Touch, being a less widely adopted mobile OS, may have fewer security researchers and resources dedicated to malware analysis. This could result in a reduced availability of comprehensive malware detection tools and a slower response to emerging threats.
5. **KaiOS**: While KaiOS has gained popularity as a feature phone OS, it may still face limitations in malware analysis due to its unique architecture and restrictions on app development. The limited app ecosystem and strict app vetting process can act as deterrents for potential malware developers, but they can also limit the scope of malware analysis.

6. **Sirin OS**: As Sirin OS is a niche OS primarily designed for enhanced privacy and security, the availability of malware analysis resources and research may be more limited compared to mainstream OS. This can restrict the depth and breadth of malware analysis conducted for Sirin OS.
7. **HarmonyOS (Hongmeng)**: Given that HarmonyOS is a relatively new OS, the availability of comprehensive malware analysis and detection tools specifically tailored for HarmonyOS may still be in the early stages. The limited time since its release may result in a smaller dataset for malware analysis and fewer established best practices for mitigating HarmonyOS-specific threats.

5.4 Mitigating Attacks: The Current Solutions

Every operating system is concerned about security flaws since they can put user data and individual devices at risk. Each OS developer applies different security measures and embeds security features within their systems to overcome these issues.

This section looks at the mapping of security weaknesses to the OS-secured elements in seven different operating systems such as Symbian, Tizen OS, Sailfish OS, Ubuntu Touch, KaiOS, Sirin OS, and HarmonyOS. By examining the relationship between vulnerabilities and secured components, we can gain insight into the effectiveness of the security mechanisms used by each OS. This analysis helps in comprehension of the benefits and drawbacks of different OS for reducing security threats, thereby advancing secure computing environments (Table 5.4).

Table 5.4 Mapping between security vulnerabilities and their secured element

Mobile OS	Security vulnerabilities	Secured element
Symbian	Lack of stability when presented with corrupted or nonstandard system files, vulnerable to the overwritten system, applications with corrupted ones, no concept of roles or users, and no access controls in the file systems [9–11]	Kernel separation, preventing loading device drivers in the kernel, disallowing overriding of ROM-based plug-ins
Tizen	Lack of third-party app support, the potential for security vulnerabilities in the web runtime	Secure boot, SELinux, app sandboxing
Sailfish	Limited app ecosystem, the potential for security vulnerabilities in third-party apps	Encrypted data storage, device encryption, user control over permissions
Ubuntu Touch	Limited app availability, the potential for security vulnerabilities in Libertine (privilege escalation and code injection) [16]	App confinement, secure boot, system-wide updates
Kai	Limited functionality compared to full-fledged operating systems	App verification, security updates, user data encryption
Sirin	Potential security vulnerabilities in Android subsystems, reliance on the Android app ecosystem	Secure element for cryptocurrency transactions, hardware-based security features
Harmony (Hongmeng)	Potential vulnerabilities in the app ecosystem, lack of third-party app support	Trusted execution environment, sandboxing, device virtualization

5.5 Utilizing Services: The Trend Now

With the advent of modern mobile OSs, such as Android and iOS, the other mobile OSs that we described in this chapter were eventually discontinued. The following are some utilizing cases and applications of those mobile OSs.

- Tizen OS-based Samsung Smart TV
- All Samsung Smart TVs are built-in with Tizen OS. Tizen is a Linux-based open-source OS developed by Samsung and Intel. Since Tizen OS is open source, it is readily available for all developers to build an app that is optimized not only for smart TV but other connected devices as well.
- Alternative to Android OS
- Although Ubuntu Touch OS was abandoned and is no longer used, it was picked up by the community, which formed UBPorts. Like the similarly abandoned Unity, UBPorts continues the work started by Canonical, and now the operating system (Meizu Pro 5 in 2016) is available on a much bigger collection of phones. It is available as a mobile app. Once installed, it is built into the mobile OS. The Ubuntu Touch mobile operating system gives the user a brand-new way of using the phone [12].
- Low-cost devices' operating system
- KaiOS is used for extremely low-cost devices in developing markets such as India, Nigeria, and Indonesia. KaiOS is a "smart feature phone" OS designed to reach the next billion people who have not gotten on the Internet yet. When there is a need for a phone to take when life is meant to be lived without the distraction of an expensive smartphone sending hundreds of notifications, a KaiOS device is affordable and connected enough to do the job [13].
- Cryptocurrency in a mobile environment
- Sirin Labs' proprietary OS (Sirin OS) gives users an easy-to-use Android experience with the Play Store and the apps it contains. Sirin OS is secure enough for storing and using cryptocurrency in a mobile environment. The main benefit of Sirin OS is the enhanced security of the entire device and the built-in cold wallet accessible via the Safe Screen. Users will have a familiar experience, along with an extension beyond the Android OS, to ensure blockchain and wallet security. These factors will enable its device to be considered in the mass adoption of blockchain technology [14].
- Smart distribution
- Huawei's HarmonyOS is a distributed operating system to collaborate and interconnects with multiple smart devices on the Internet of Things (IoT) ecosystem. It is designed to be a one-stop shop for all platforms. That means a user can expect to find a matching set of features, apps, and support across the board. For users, HarmonyOS forms a super device that connects various smart devices and enables them to share data with each other and facilitates cross-device collaboration. For developers, HarmonyOS abstracts hardware capabilities, which allow you to develop an app only once and deploy it across a variety of devices. Your app will cover the most users with the least investment [15].

5.6 Chapter Summary

The alternative mobile OSs covered in this chapter, namely Symbian, Tizen OS, Sailfish OS, Ubuntu Touch, KaiOS, Sirin OS, and Harmony OS, serve as excellent examples of how the field of smartphones or mobile computing is constantly changing. To meet various user needs and device specifications, each OS brings its own special features, target market, and design philosophies.

Even though certain OSs have encountered difficulties, such as constrained app ecosystems and security flaws, they continue to develop and adapt to satisfy user needs. These OS improvements have opened the door for mobile technology innovation.

It is intriguing to think about how these OS will continue to influence mobile computing in the future as the industry develops.

> By reading this chapter, you can answer the following questions:
> Can you explain the history of each of the six (**seven?**) alternative OS?
> What factors led to the creation of each new OS?
> How does each niche OS differ from other common OSs?
> What are the security vulnerabilities of each OS?
> Which OS is the most secure?

References

1. Muthukumaran, D., Anuj, S., Schiffman, J., Jung, B. M., & Jaeger, T. (2008) Measuring integrity on mobile phone systems. In *The 13th ACM symposium on Access control models and technologies*. Estes Park.
2. Kumar, R. (2023). *Symbian Os*. Accessed 2023, from https://www.codingninjas.com/codestudio/library/symbian-os#:~:text=in%20the%20article.-,Symbian%20OS%20architecture,kernel%20also%20performs%20preemptive%20multitasking
3. All About Symbian, http://www.allaboutsymbian.com/
4. https://developer.tizen.org/zh-hans/tizen-architecture
5. https://sailfishos.org/content/uploads/2018/11/Sailfish_Architecture.jpg
6. https://phone.docs.ubuntu.com/en/devices/porting-new-device
7. https://developer.kaiostech.com/docs/
8. https://www.huaweicentral.com/huawei-publishes-january-2023-harmonyos-security-patch-details/
9. Li, B., Reshetova, E., & Aura, T. (2010). *Symbian OS platform security model. For a complete list of all USENIX & USENIX co-sponsored events*, see http://www.usenix.org/events
10. Gharibi, W. (2012). *Symbian vulnerability and mobile threats*. arXiv preprint arXiv:1201.0945.
11. Haas, J. D. (2005). *Symbian phone Security*. Black Hat Europe. Accessed 2023, from https://www.blackhat.com/presentations/bh-europe-05/BH_EU_05-deHaas.pdf
12. Cawley, C. (2022). How to Install Ubuntu Touch on Your Mobile Phone. *Make use of*. Accessed 2023, from https://www.makeuseof.com/how-to-install-ubuntu-touch-on-your-mobile-phone/

References

13. Zahran, O. (2020). KaiOS: The most important operating system. *The Startup*. Accessed 2023, from https://medium.com/swlh/kaios-the-most-important-operating-system-2d92644959a4
14. *SIRIN OS, Our Ecosystem. SIRIN LABS, 2023*. Accessed 2023, from https://sirinlabs.com/sirin-os/
15. Ozhayta, A. M. (2020). *HarmonyOS*. HUAWEI Developers . Accessed 2023, from https://medium.com/huawei-developers/harmonyos-4bfe31c99be7
16. https://fr.m.wikipedia.org/wiki/Fichier:Architecture_Ubuntu_Touch.png

Chapter 6
Mobile Application Security

Due to the high penetration rate of mobile applications in users' lives, any security threats of mobile apps directly affect users' activities. In most cases, attackers steal personal information and track users. Complicating matters further is the fact that mobile devices have limited computational power and a restricted user interface, making it easier for attackers to hide their malicious activities. In this chapter, we discuss mobile application (mobile apps) security and related important factors, along with a set of recommendations for mobile users and App developers.

6.1 Application Security Threats

Four primary methodologies are employed in mobile application development:

1. **Native mobile applications**—are crafted using the language and frameworks offered by the platform owner and operate directly on the device's OS, such as iOS or Android.
2. **Cross-platform native mobile applications**—can be developed with various languages and frameworks but are compiled into a native app running directly on the device's OS.
3. **Hybrid mobile applications**—built with standard web technologies such as JavaScript, CSS, and HTML5, these apps are packed as app installation packages. Unlike native apps, hybrid apps operate on a "web container" that supplies a browser runtime and a bridge for native device APIs, typically via Apache Cordova.
4. **Progressive web applications (PWAs)**—offer a unique approach to conventional mobile app development. PWAs are web applications that use certain browser capabilities to offer an "app-like" user experience, forgoing the need for app store delivery and installations.

A security threat is defined as any action that takes advantage of security weaknesses in a system and has a negative impact on it. As mobile applications become a reality, a growing number of ubiquitous mobile devices have raised the number of security threats.

Basically, mobile application security should not simply focus on data and applications. Mobile platforms are used in various new settings and impact users in ways that could never apply to a PC. An attacker could compromise systems connected to mobile devices through vulnerabilities identified at any point. The rise of mobile botnets is a characteristic example of such a case. Furthermore, as we anticipate a shift from mobile platforms to wearable devices (smart watches, glasses, and the like), there are even more reasons to worry. Thus, we argue that it is much more interesting and challenging for security communities to work on mobile application security—not only because it is an emerging topic but also because it could have a much greater impact on how we think about system security research.

Normally, mobile app attacks may occur in all layers, from the application layer to the physical layer. For example, at the application layer level, a malicious attack can be added along the communication link to generate fake messages and data in order to attack ongoing communication and increase data collision. The attack in the transport layer happens through sending unlimited connection requests in order to minimize the node's energy and exhaust its resources, and this leads to denial of service. Other attacks can occur in a network layer in several forms, such as spoofing, sinkhole, flooding, and replay attacks in order to create and send fake messages or cause congestion in the network. Jamming attacks at the data link layer can cause loss of signals and data and destroy the channel and increase interference. At the physical layer level, the attacker can allow unauthorized nodes to access the network and damage it. Mobile security threats can be classified into technical and nontechnical resource threats. Technical threats were categorized into three types of threats: infrastructure threats, technical, operational threats, and system data management threats. In comparison, nontechnical threats were classified into environmental threats and governmental threats. Authentication attacks, side-channel attacks, privacy leakage, cloud malware injection attacks, Denial of Service (DoS) attacks, and service manipulation are some examples of these security threats.

In the view of the organization, mobile apps have had a transformative effect. Through ever-increasing functionality, ubiquitous connectivity, and faster access to mission-critical information, mobile apps continue to provide unprecedented support for facilitating organizational objectives. Despite their utility, these apps can pose serious security risks to an organization and its users due to vulnerabilities that may exist within their software [1]. Such vulnerabilities may be exploited to steal information, control a user's device, deplete hardware resources, or result in unexpected app or device behavior.

Data confidentiality is considered one of the fundamental problems in mobile applications. In the context of mobile platforms, if the user needs to access data, authorization should be taken first in order to prevent attackers from accessing sensitive data stored on mobile devices. To achieve that, there is a need to focus on two important aspects: (1) authorization and access control and (2) identity authentication. Mobile devices and applications need to be able to verify the user or device identity is authorized to access the data or not. Where an authorization mechanism helps mobile devices and applications to identify if the mobile users or devices are permitted to access data or services. Access control mechanism also ensures that attackers have access to the resources of the system. This will establish a secure connection between mobile devices and, thus, the transition of data between users in a safe way. Another important issue that should be considered in mobile devices and applications is identity authentication. In fact, this issue is very critical in the mobile environment because a huge number of users and devices need to authenticate each other in a trustable way in a secure manner. Privacy is an important issue in mobile devices and applications for users, organizations, and governments. In the context of mobile platforms, mobile devices are connected, and sensitive data is shared and exchanged over the internet; this makes user privacy a sensitive topic in the mobile domain. Protecting the privacy of users' data in mobile devices from cyberattacks is still a hot topic for many researchers and needs to address.

6.2 Application Vulnerabilities

App vulnerabilities, often a result of design flaws or programming mistakes, are prevalent in the app marketplace. These errors might be intentional or unintentional, with some developers prioritizing functionality over security to minimize costs and accelerate market release.

Mobile operating system vendors such as Android and iOS host commercial app stores where they review apps for issues such as malware, user information collection without consent, objectionable content, and performance impact before offering them in their marketplace. However, the depth and nature of these reviews are not transparent to consumers or government bodies. These marketplaces cater to billions globally and focus on consumer and brand protection. Consequently, enterprises, federal agencies, and regulated industries planning to utilize consumer apps must assess risk-based decisions regarding app procurement based on their unique security, privacy, policy requirements, and risk tolerance.

The risk level of vulnerabilities varies. Apps that handle sensitive data such as precise geolocation details, personal health metrics, or personally identifiable information (PII) are deemed higher risk. Moreover, apps relying on wireless network technologies for data transmission, such as Wi-Fi, cellular, or Bluetooth, can also pose high risks, as these technologies can be exploited for remote attacks. Nonetheless, even low-risk apps can cause substantial damage if exploited—for instance, public safety app vulnerabilities could lead to loss of life.

To reduce security risks associated with mobile apps, organizations should implement a software assurance process. This provides confidence that the software is free from vulnerabilities, intentionally designed or accidentally inserted, and operates as intended. This document outlines such a process specifically for mobile applications, termed an "app vetting process."

6.3 Information Sensitivity

Although there is a wide range of mobile applications (e.g., games, lifestyles, financial, utility, and education), the sensitivity of the information given to each app is different. Users have different privacy and security perceptions toward information disclosure based on the type of information shared with each app. That is, individuals are more concerned about releasing information to financial apps than fitness or game apps. Users have different levels of concern.

When sharing more sensitive information, finding that when users are requested to provide more sensitive information, they perceive a higher risk and have lower intentions to disclose the requested information. Despite the importance of information sensitivity in privacy research, prior research on mobile apps does not consider the effect of information sensitivity on privacy-related relationships. In general, sharing more sensitive information with mobile apps causes users to be more concerned because if the shared information is disclosed to third parties, users must deal with consequences that can disrupt their daily lives. When users are required to provide more sensitive information, they become more conscious of the consequences of disclosing information to mobile apps. When a mobile app requests access to the mobile device's location service to track users' locations, users focus more attention on giving such permission. Thus, users with the same perceptions of privacy risk perceive mobile apps that request more sensitive information to have lower security because they expect these apps to be more secure than apps requesting less sensitive information. For instance, users expect financial apps to be more secure than games, but when users perceive the same level of risk with financial apps and games, they perceive financial apps to have lower security. With the same logic, when users are required to provide more sensitive information, the presence and understanding of a privacy policy become more important, and users consider the privacy policy more. Hence, the effect of a privacy policy on the risk and security perceptions of mobile apps becomes stronger when users need to provide more sensitive information. Prior privacy studies have considered the moderating role of information sensitivity in privacy-related relationships and found that perceptions pertaining to information disclosure change when users disclose more sensitive information [2].

6.4 Semantic Security

Data transfer happens intensively in the mobile communication environment, such as location sharing, navigation, critical data inquiry, and so on. The substance of security, namely data and information security, is the most critical source of the security issue. The back end of the mobile network or app provider shares the analogous structure and technology as opposed to that of the conventional network; as a result, the substance or data security issues that happen at the back end of the system are quite similar to those in the conventional network. For example, Mukherjea and Sougata (2017) discussed data encryption, role-based access control, and identity certificate by use of trust profiling in the context of mobile privacy data inquiry and transmission. Within the context of mobile app providers, data is stored or transmitted in and between mobile devices and relevant systems. Security issues are related to the actions such as unauthenticated visiting, storage, manipulation, or data abuse. In addition, the risks from the providers of the mobile apps cloud service are also rising due to typical events such as data breaches or ransomware attacks [1].

6.5 Infrastructure-Related Security

The infrastructure system of the mobile network/application provider may be physically damaged. For instance, the signal receiver module of the bulletin board system (BBS), a computer/application dedicated to the sharing or exchange of messages or files on the network, is invalid; the entire BSS can no longer provide services. When BSS is down, a local mobile network within a certain area is out of function. It is noted that a particular BSS is only one of the nodes of the whole system, and each node may be dysfunctional. Therefore, the greater number of nodes would imply a high likelihood of failures of the system.

In addition to security issues, privacy issues are possible when a mobile app provider's infrastructure is compromised. There are two situations regarding the loss of privacy information. The first situation is that a piece of a user's private information on the provider side has been hacked. The second situation is that a piece of infrastructure over which a piece of privacy information runs has been damaged/compromised. The damaged infrastructure may carry privacy information, and there are two further cases: Case (1) the damaged infrastructure is left to the outside entity that should be anonymous to this information, and Case (2) the damaged infrastructure is lost, and the owner of this information wants to get back this information. Specifically, the problem with this case can be addressed by the design principle for information systems called the "derived information" mechanism, which may also be called redundancy data.

Malfunctions or inefficient security control of the user's mobile OS may lead to information leakage and loss of important personal information that might further result in privacy violations. For instance, side-channel attacks are common when a common resource (e.g., memory) is shared between several mobile applications operated by different providers [1].

6.6 Privacy Awareness

One of the reasons why mobile users have different privacy concerns about the same phenomena is their varying levels of privacy awareness. For example, individuals aware that some websites and applications may collect their personal information without permission have more privacy concerns than those unaware of such malicious activities. Privacy awareness refers to the degree to which a person is aware of information privacy practices in general and the use of this by mobile apps. To illustrate, an individual's privacy awareness is increased by reading and watching the news, reading books and magazines, and hearing about privacy issues from friends or other people within their social bubble. Privacy awareness can influence an individual's attitude and perceptions toward mobile apps. Privacy awareness exerts an influence on privacy-related perceptions because an individual's awareness stimulates their protective behaviors. The difference in users' levels of privacy awareness leads to different privacy and security perceptions of mobile apps. Therefore, privacy awareness increases or decreases the strength of the relationship between privacy-related concerns and security. For example, users with higher privacy awareness perceive mobile apps to have lower security because they have read or heard more about unauthorized information disclosure incidents and know that such incidents frequently transpire with less secure mobile apps. Similarly, mobile users with a higher level of privacy awareness assume that privacy policy statements incorporating privacy and security practices do not necessarily indicate a higher level of mobile app security due to the prevalence of privacy and security incidents despite the providers' protection efforts [2].

6.7 The Best Practices for Mobile Users on App Security

It is well-known that our technology usage can expose us to viruses and cyberattacks if we do not take proper security measures. As discussed in our previous chapters, mobile devices often lack a comprehensive user manual that educates users on mobile security. Additionally, threats in the digital landscape are constantly evolving and adapting to exploit our habits. We have compiled a list of crucial security practices to ensure your mobile device usage is secure.

- **Avoid saving passwords**
- Modern mobile apps can access various system resources, sometimes without explicit user permission. Despite strict separation enforced by the OS, apps can still access private information from other apps through side-channel attacks. For

instance, a zero-permission app might indirectly infer keystrokes from an accelerometer or gyroscope output, potentially learning sensitive data like passwords.
- Storing passwords, a practice not recommended due to security issues, is particularly crucial when dealing with apps for online banking or healthcare. Protective measures such as encryption or passcode-based access control should be implemented when password storage is necessary. Modern smartphone OSs provide such protections; for example, iOS devices auto-encrypt content once a passcode is configured, and Android users can enable data encryption via the security settings.
- **Avoid downloading malicious and fake apps**
- Smartphones, equipped with a range of hardware sensors, have access to sensitive user information, making them attractive targets for malicious apps [3]. The Global Mobile Security Report 2022 [4] revealed that 22% of mobile devices house at least one app that unnecessarily extracts user data. Mobile spying comes in two forms—the more common mass data collection by apps for advertising purposes and the rarer advanced hacking techniques targeting specific individuals, such as the Pegasus spyware. Avoiding downloads from unknown app stores and unauthorized providers is recommended.
- Over 700 app stores provide copies of official apps. In 2020, three third-party stores surpassed Google Play and the App Store regarding new apps uploaded. Modified applications (MODs) or altered versions of original apps often contain malicious code, enabling data extraction, user spying, and unwanted ad display [4].
- **Be cautious about public Wi-Fi networks**
- Be mindful of potential security threats when using public Wi-Fi networks in cafes or restaurants. These networks can be prone to Man-in-the-Middle (MitM) attacks, where an attacker manages the network and sifts through the traffic in search of vulnerabilities. Consider using security applications that encrypt your mobile network traffic to protect your data.
- Always establish a secure connection before engaging in any activities over the public network. This encrypted data transmission safeguards your credentials against interception attempts. If unencrypted, your login details for online services, such as banking apps, could be at risk. It is advisable to use your mobile data network instead of public Wi-Fi to enhance security.
- **Be restrictive with your mobile app permissions**
- Mobile app permissions protect all sensitive information stored on a given device. These permissions should only ask for access to information necessary for the app to function properly. Access to the user's location, camera, microphone, contact list, system files, or phone's IMEI number should be restrictedly granted. Permission requests are there to protect sensitive information that is on the device. Permission requests should be used only to access information necessary for the app's functioning. There are a variety of permissions that differ from app to app. Google and Apple have guidelines that let users decide which permissions to allow or revoke [5].

- **Be aware of social engineering**
- Social engineering, traditionally associated with tricking computer users into sharing passwords or installing malicious software, is also a considerable risk for mobile users. You should be cautious of unexpected calls or emails prompting you to install an application or enter credentials on your smartphone. Always scrutinize the source and only proceed if it is genuinely trustworthy.
- **Update your smartphone OS**
- It is imperative to regularly update your smartphone's operating system (OS). These updates not only augment the device's performance and compatibility with newer applications but also elevate its security measures. The updates contain patches that protect your device against cyber threats and malware. Hence, maintaining an updated OS is a crucial step for efficient and secure smartphone usage.
- **Turn off your mobile's Bluetooth**
- Bluetooth can be a potential gateway for hackers to gain access to your smartphone, as it enables device interaction. Ensure Bluetooth is disabled if it is not actively needed, especially in public wireless environments, to minimize potential security risks.
- **Update your applications**
- Always keep your smartphone applications current by frequently updating them or moving to the most recent versions. Ignoring these updates can leave your applications vulnerable to the latest security risks. Unfortunately, some app developers do not routinely update their apps or acknowledge smartphone OS updates, creating a potential security gap against emerging threats.

6.8 The Best Practices for Smartphone App Developers

Unique to the mobile landscape, security challenges faced by apps change significantly after download. Users gain access to source code, potentially revealing weak spots, and a jailbroken device can leak app data, opening doors to other apps or potentially harmful actors. Unlike desktops with typical antivirus protections, mobile updates rely on user discretion, increasing the risk. To counter these, additional security measures are required to limit exploitability. Here, we outline the following essential practices for smartphone app developers [6–10]:

- **Protect the app code encryption**
- Encryption is the process of converting data from plain text to ciphertext, which cannot be read without the decryption key. Encryption can either be symmetric, which means the same key is used for encrypting and decrypting, or asymmetric, which means a different key is used for decryption.
- Encryption should be applied to data at rest, as well as in transit, using an SSL or VPN tunnel. To protect data at rest, you should be sure that you are encrypting

data at the file or database level. Operating systems provide sandboxes for each application where unstructured data may be stored. These sandboxes are designed to section off each application's data from the rest of the applications on the device; however, not all application data is stored here. It can also be stored in external storage, making it vulnerable to attack.

- Another important thing to consider with encryption is proper key management. It does not matter that you have locked the door if you leave the key under the mat. You can use a key management system to keep keys separate from the data they are meant to protect. You will need to develop policies for things such as the length of time a key is authorized to be in use, key size based on best practices for the type of data and type of key, and retirement of unneeded keys.
- **Code obfuscation**
- Code obfuscation is the process of making source code difficult to decipher to prevent unauthorized use. Obfuscation alters the code without changing the end result, leaving it functionally equivalent to the original code. This differs from encryption in that there is no key for obfuscation. Obfuscation just makes it harder to read the data instead of changing the data itself.

There are several ways to do this:

- **Name obfuscation:** Changing the names of functions, classes, and variables. This is helpful, but by itself is usually not enough, as a hacker will still be able to figure out the patterns and understand which functions do what, even if they go by different names.
- **Control flow flattening:** Taking the basic blocks of the source code and putting them into a loop, making the program flow much less obvious.
- **Arithmetic obfuscation:** Taking arithmetic calculations and replacing them with much harder-to-understand mathematical equivalents.
- **Code virtualization:** Translating the code into self-defined bytecode that can only be understood by a virtual machine.

Code obfuscation is helpful for preventing hackers from reverse engineering your source code, which can result in the loss of sensitive data or the creation of a rogue app. Rogue apps are apps created by a threat actor to resemble the app of a known brand with the goal of tricking users into downloading the malicious app instead.

- **Secure your APIs**

- While APIs provide enormous convenience, they are also a huge security risk. APIs generally use an API key to verify that what is making the call to the API server is a legitimate instance of your mobile app. But what happens when an attacker has that API key? Without additional methods of identity verification and the ability to detect patterns indicative of abuse, such as calls by bots, it becomes increasingly difficult to differentiate real user API calls from malicious ones.

- There are a number of things you can do to reinforce your API security. For one, you should ensure robust encryption between the API client and server to reduce the impact of man-in-the-middle attacks. You also need to be aware of where you are storing your API keys. When API keys are stored in a mobile application, threat actors are able to use reverse engineering tools to access that information, hence the need for techniques such as code obfuscation. Because API keys are so discoverable, securing your APIs will depend on being able to effectively authenticate the entity making the request.
- **Strong authentication policies**
- Strong credentials are a must for both web and mobile application development. For mobile apps, you can choose to either have a native login flow, which means the user enters their credentials within the app, or a web-based login flow, where they are directed to a web browser to log in. Native login flows provide a better user experience but are generally thought to be less secure. Hypermedia authentication APIs are a solution now popping up to bridge this gap and provide the best of both worlds. Hypermedia authentication APIs interact with the authorization server directly without the need for an intermediary such as the browser window.
- Regardless of how the user enters their credentials, your app should enforce some type of password policy to ensure a strong password is used, and it should not store the access and refresh tokens anywhere except secure storage (such as the iOS keychain or Android Keystore). You can also take advantage of secondary forms of authentication such as biometric methods like facial recognition.
- You should also ensure that you are using proper session handling. Mobile applications tend to have longer sessions than desktop apps. Long sessions can be beneficial for businesses that want to encourage ease of use for customers; however, they can pose significant security risks. Ensure that your tokens are long, complex, and random to reduce the attackers' ability to brute-force them. Session invalidation should occur on the mobile app as well as on the server side to prevent HTTP manipulation tool attacks.
- **Secure the data-in-transit**
- Critical user data being sent between client and server must be shielded from theft or privacy leaks. Employing an SSL or VPN tunnel for rigorous protection is strongly suggested.
- **File-level and database encryption—make provisions for data security**
- Mobile applications typically store unstructured data in the device's local file system or database, creating potential vulnerabilities as this data is not always effectively encrypted. To address these security gaps, it is crucial to use tools such as SQLite Database Encryption Modules or apply file-level encryption across different platforms, ensuring a more secure sandbox environment.
- **Have a solid API strategy**
- APIs facilitate data exchange between apps, users, and cloud services; thus, their security is critical. If your app utilizes third-party APIs, ensure this only access necessary parts of your app, thereby minimizing vulnerabilities.
- **Secure the back end**

6.8 The Best Practices for Smartphone App Developers

- Most mobile apps operate on a client–server system, necessitating protective measures against back-end server attacks. Do not assume your app is the only one accessing your APIs. Validate all APIs in line with your target mobile platform, considering that API authentication and transport can vary across platforms.
- **Store private data within internal storage**
- Hold user-specific sensitive data within the device's protected internal storage. No permissions are required for your app to access these files, which are inaccessible to other apps. When your app is uninstalled, these files are automatically deleted, ensuring security.
- **Follow secure coding practices**
- Just like web apps, mobile apps should adhere to secure coding best practices from the design stage and throughout the development lifecycle. Although open-source code bases facilitate faster development, they can pose serious security risks. A combination of Static (SAST) and Dynamic Application Security Testing (DAST) can help uncover vulnerabilities in both the development and postproduction stages.

  Some additional secure coding recommendations are as follows:

- Regularly perform mobile application security testing on your code for bugs and fix them.
- Keep your code agile so that it is possible to perform a real-time update at the user end after a breach to fix it.
- Take a default deny approach to data and enforce the principle of least privilege.
- Validate inputs from all untrusted data sources.
- Sanitize data sent to other systems.
- Implement a system for monitoring and logging.
- Keep it simple. Complex designs leave more room for vulnerabilities. The more things the app can do that it does not need to do, the more room there is for attackers to exploit these unnecessary functions.
- Conduct code audits and tests to see that the app's authentication and authorization procedures have no loopholes.
- Check access controls to detect data security issues before they become large-scale problems.
- Use a gateway to protect your API.
- Use operating system emulators to see how your app would perform in a simulated environment.
- While creating your data storage systems, keep in mind that no sensitive data should be shared with:
  - the application log
  - third parties
  - the keyboard cache
  - the IPC mechanism
  - the user's device during interaction

- Use data analytics to perform dynamic analysis and note how, when, and where data moves on your application.
- Including multifactor authentication by mandating the usage of a one-time password (OTP) in addition to the normal password.
- Add an additional security layer through biometric authentication that needs fingerprints or retina scans.
- **Hire a mobile app security expert**
- As data breaches become more common, app security has become a paramount concern. To ensure your app's safety and data integrity, consider hiring a security specialist. They can provide expert guidance to fortify your app against potential security threats.
- **Empower your users**
- Even the most secure apps rely on users' behavior for their security. As a developer, it is important to educate your users about online safety practices. This not only helps protect them but it also enhances the overall security of your application.

6.9 Chapter Summary

This chapter discusses some security issues with mobile applications and recommends solutions to avoid those issues. Mobile applications are divided into devices' inbuilt apps, developed and installed by the manufacturers, and the additional apps developed by the service providers or developers. It is recommended that mobile apps should be downloaded from authorized and reliable app stores and check all permissions granted to them. Updating the smartphone OSs and installing the last security patches are recommended for all mobile OS types. To avoid mobile apps harmful to the company and its workforce, MAM tools are suggested to detect threatening apps. Monitoring mobile apps is necessary to detect malicious activities and malware. Similar recommendations are discussed for mobile app developers. It is recommended that app developers use clear authentication policies, solid and secure API strategy, and follow secure coding.

> By reading this chapter, you can answer the following questions:
> What types of security threats are caused by mobile apps?
> How is a mobile user's privacy compromised by malicious mobile apps?
> What are the possible security vulnerabilities of mobile apps?
> What is the best protection for mobile users' in-app security?
> How can mobile app developers provide secure and reliable apps?

References

1. Xu, M. (2019). *A system perspective to privacy, security and resilience in mobile applications*. Ph.D. diss., University of Saskatchewan.
2. Balapour, A., Nikkhah, H. R., & Sabherwal, R. (2020). Mobile application security: Role of perceived privacy as the predictor of security perceptions. *International Journal of Information Management, 52*, 102063.
3. Osman, T., Mannan, M., Hengartner, U., & Youssef, A. (2020). Securing applications against side-channel attacks through resource access veto. *Digital Threats: Research and Practice, 1*(4), 1–29.
4. Pradeo. (2022). *Global Mobile Security Report*. Accessed 2023, from https://marketplace.soti.net/images/uploads/documents/pdf/1690/mobile-security-report-2022.pdf
5. Inapptics. (2018). *3 Best practices for in-app permissions*. https://blog.prototypr.io/3-best-practices-for-in-app-permissions-dce7d36544a4
6. van der Linden, D., Anthonysamy, P., Nuseibeh, B., Tun, T. T., Petre, M., Levine, M., Towse, J., & Rashid, A. (2020). Schrödinger's security: Opening the box on app developers' security rationale. In *Proceedings of the ACM/IEEE 42nd international conference on software engineering (ICSE '20)* (pp. 149–160). Association for Computing Machinery. https://doi.org/10.1145/3377811.3380394
7. Weir, C., Rashid, A., & Noble, J. (2016). *How to improve the security skills of mobile app developers?: Comparing and contrasting expert views*.
8. Hatamian, M. (2020). Engineering privacy in smartphone apps: A technical guideline catalog for app developers. *IEEE Access, 8*, 35429–35445. https://doi.org/10.1109/ACCESS.2020.2974911
9. Chen, E. Y., Pei, Y., Chen, S., Tian, Y., Kotcher, R., & Tague, P. (2014). OAuth Demystified for mobile application developers. In *Proceedings of the 2014 ACM SIGSAC Conference on Computer and Communications Security (CCS '14)* (pp. 892–903). Association for Computing Machinery. https://doi.org/10.1145/2660267.2660323
10. He, B., Xu, H., Jin, L., Guo, G., Chen, Y., & Weng, G. (2018). An investigation into android in-app ad practice: Implications for app developers. In *IEEE INFOCOM 2018 - IEEE conference on computer communications*, Honolulu, HI, pp. 2465–2473. doi: https://doi.org/10.1109/INFOCOM.2018.8486010.

Chapter 7
The Best Security Practices

This chapter provides a comprehensive list of the essential practices that we highly recommend for all smartphone users, whether for business or personal purposes. Implementing these practices is more crucial than ever due to the increasing prevalence of cybersecurity crimes [1–7].

- **Choose the right mobile OS for your risk tolerance**
- Choosing between Android and iOS involves an evaluation of security and privacy aspects. While Android and Windows phones provide open-source compatibility, cost-effectiveness, and diverse apps, iOS devices typically offer superior security due to prompt updates and comprehensive privacy protections.
- **Implement authentication and access management**
- Implementing multi-factor authentication (MFA) enhances security by demanding users to present two out of three authentication elements: something they possess (e.g., mobile device), something they know (password, PIN, OTP), and something inherent to them (fingerprint, face ID). Access management assigns roles and security levels to users, with the flexibility to modify parameters based on risk factors and device trustworthiness, deciding the necessity of MFA.
- **Regularly update your mobile devices and apps**
- OS and app updates frequently carry security patches addressing known vulnerabilities. Given that these updates are not always automatic, maintaining your device's security demands regular updates of your OS and installed applications. Ignoring updates leaves your device's data prone to risks. Enabling automatic updates and regularly checking for new ones is a best practice for maintaining device security.
- **Avoid public Wi-Fi**
- Public Wi-Fi, despite its convenience, may expose your device and data to risks. Fake Wi-Fi networks set up by hackers or data interception on public networks can compromise your confidential information. Avoid public Wi-Fi whenever possible. If Wi-Fi access is necessary, connect through a VPN for improved security.

© The Author(s), under exclusive license to Springer Nature
Switzerland AG 2024
A. F. Abdul Kadir et al., *Understanding Cybersecurity on Smartphones*,
Progress in IS, https://doi.org/10.1007/978-3-031-48865-8_7

- **Turn off Bluetooth and Wi-Fi when not in use**
- Bluetooth and Wi-Fi can become gateways for hackers. Bluetooth, especially in discovery mode, is particularly susceptible to attacks. Thus, disable Bluetooth when not in active use. While Wi-Fi is typically more secure, turning it off when not needed is still advised.
- **Introduce password managers**
- Avoid storing passwords on unprotected apps or notes due to the associated risks. Using a password manager that provides a secure one-stop location for all your passwords, protected by a master password. Password managers also facilitate the generation of secure passwords, obviating the need for weak, easily memorable passwords. For added security, consider coupling your password manager with an MFA app.
- **Utilize a VPN**
- A Virtual Private Network (VPN) is essential for uncertain network connections. It ensures your browsing on public Wi-Fi remains private and offers added security when accessing potentially insecure sites. With affordable options available, VPNs play a pivotal role in protecting web traffic and personal data.
- **Monitor links and websites carefully**
- Be careful when clicking on links, particularly from unknown sources. Cybercriminals often use malicious links in emails or ads to compromise devices. If a link appears suspicious, avoid it. If possible, use a more secure device, such as an iOS device, for opening uncertain links.
- **Remote lock and data wipe policy**
- In a Bring Your Own Device (BYOD) environment, include policies for remote device lock and data wipe. This empowers the organization to remotely lock or erase all data if a device is lost or stolen. Despite the potential loss of personal data, this measure helps prevent unauthorized access to sensitive company information.
- **Do not jailbreak your smartphone**
- Jailbreaking your phone, an unauthorized modification, opens the door to potential malware and security threats. Even on iOS devices, jailbreaking can compromise the built-in safety features, making your device more vulnerable. If your device has been jailbroken, consider restoring the original operating system through an update or seeking help from an authorized reseller.
- **Utilize mobile device management (MDM)**
- Implementing mobile device management (MDM) can be highly beneficial. MDM allows you to remotely monitor, manage, and configure various devices your organization uses such as laptops, smartphones, and tablets. This helps in securing data stored on remote servers.
- **Don't forget the backup**
- Ensure your data is safe by maintaining regular backups, preferably on an automated cloud solution. This is vital in case of device loss or theft. For easy recovery, choose a cloud service that keeps a version history for at least 30 days such as Google's G Suite, Microsoft Office 365, or Dropbox.

- **Install an antivirus and anti-malware software**
- Safeguard your device against malicious code by installing a reliable antivirus app. Some options provide extra features such as data erasure for lost devices, caller tracking, and unsafe app identification. Clear web browsing history and delete cookies, which can store sensitive login details and be exploited by malicious actors.
- **Utilize encryption**
- To enhance mobile security, encrypt both data stored on your device and data transmitted to and from it. Utilize a VPN (Virtual Private Network) to achieve this. When connected to public Wi-Fi, refrain from sharing personal and sensitive information due to risks such as poor encryption, Man-in-the-middle attacks, data alteration, and eavesdropping.
- **Implement firewall protection**
- Protect your mobile devices from network-based threats such as malware and viruses by implementing firewalls. These network security devices prevent unauthorized access and safeguard your devices.
- **Beware of phishing scams**
- Phishing scams often arrive through emails or messages with malicious links or attachments, targeting data on mobile devices for malicious intent. Be cautious of suspicious emails, especially those promising unrealistic benefits or requesting urgent credentials. Always approach such messages with skepticism to avoid falling victim to phishing scams.
- **Be mindful of granting permissions**
- During app installations, carefully evaluate the permissions you grant. Ensure that access to contacts, gallery, camera, and authentication aligns with the app's purpose. Avoid granting unnecessary privileges, as this may raise suspicion. Always consider what level of access an app truly needs to fulfill its intended functions before granting permissions.
- **Conduct regular security audits**
- Regular security audits are crucial for identifying and addressing vulnerabilities in mobile devices. Utilize tools such as vulnerability scanners and penetration testing to conduct these audits. Businesses should update their security policies to comply with industry standards and regulations.
- **Educate your employees**
- Comprehensive employee training on mobile device security is vital in preventing data breaches. Teach best practices such as avoiding public Wi-Fi and not sharing access passwords. Stress the importance of data security and how to report security incidents to the IT department effectively.

7.1 Chapter Summary

This chapter recommends that mobile device users implement a set of best practices for enhanced device security. These practices include the following: choosing the right mobile operating system based on security considerations; implementing user authentication with options such as passwords, PINs, and biometrics; regularly updating mobile devices and apps to address vulnerabilities; avoiding public Wi-Fi networks; and disabling Bluetooth and Wi-Fi when not in use. Additionally, utilizing password managers and VPNs for secure connections, exercising caution when clicking on links or opening websites, and implementing remote lock and data wipe policies in case of device loss or theft are crucial steps.

Furthermore, it is essential to refrain from jailbreaking smartphones, utilize mobile device management (MDM) for remote monitoring and configuration, regularly backup data, install antivirus and anti-malware software, employ encryption for data protection, implement firewall protection, be cautious of phishing scams, and carefully grant permissions to apps. Alongside these technical measures, educating employees about mobile device security best practices, ensuring their awareness of potential risks, and correctly reporting security incidents to the IT department are vital for preventing data breaches. By following these practices, smartphone users can significantly enhance their mobile device security and reduce the risk of cyber threats and unauthorized access.

> By reading this chapter you can answer the following questions:
> What is the best OS for your situation?
> How can you implement user authentication on your mobile?
> Why is it important to turn off your Bluetooth and Wi-Fi when they are not in use?
> What is a password manager and why is it helpful to have?
> What is a remote lock and data wipe policy?
> Why do we need regular device data backups?
> How can we set up firewall protection?

References

1. Valcke, J. (2016). Best practices in mobile security. *Biometric Technology Today, 2016*(3), 9–11.
2. Wibowo, K., & Ali, A. (2016). Mobile security: Suggested security practices for malware threats. *Competition Forum, 14*(1). American Society for Competitiveness.
3. Oh, T., Stackpole, B., Cummins, E., Gonzalez, C., Ramachandran, R., & Lim, S. (2012). Best security practices for Android, blackberry, and iOS. *2012 The First IEEE Workshop on Enabling Technologies for Smartphone and Internet of Things (ETSIoT)* (pp. 42–47). Seoul.
4. Kostic, S. Mobile security in 2022: What to expect & how to prepare. *Security Magazine*, January 2022.

5. Costello, M., 10 Mobile security best practices to consider in 2023, *Solutions Review*, March 2023.
6. Bhattacharya, P., Yang, L., Guo, M., Qian, K., & Yang, M. (2014). Learning mobile security with labware. *IEEE Security and Privacy, 12*(1), 69–72. https://doi.org/10.1109/MSP.2014.6
7. Sheila, M., Faizal, M. A., & Shahrin, S. (2015). Dimension of mobile security model: Mobile user security threats and awareness. *International Journal of Mobile Learning and Organisation, 9*(1), 66–85.

Chapter 8
Conclusion

This book has covered the fundamental elements necessary to understand the security issues, vulnerabilities, and countermeasures for ten different mobile operating systems (OSs): Android, iPhone OS (iOS), Mobile Windows, Symbian, Tizen, Sailfish, Ubuntu Touch, KaiOS, Sirin, and HarmonyOS. We have also outlined several recommendations and best security practices for mobile developers and device users.

It has been made strikingly clear from the research that every mobile OS has its own unique challenges, and every mobile OS tries to make use of the available cutting-edge security techniques in order to safeguard user information and reduce any potential security risks. The continued development of these operating systems and the corresponding efforts to solve security issues serve as a strong reminder of exactly how crucial it is for commonly used mobile systems to strike a balance between innovation and strong security procedures in the constantly shifting world of technology.

The research also provides valuable insights into the challenges, innovations, and security considerations of each OS:

- **Android**, as the most widely used mobile OS, faces the challenge of maintaining security across a diverse range of devices and a vast ecosystem of apps. Its open-source nature has led to both innovation and security concerns, constantly focusing on improving security measures and mitigating vulnerabilities.
- **iPhone OS (iOS)** stands out for its tightly controlled ecosystem, providing a high level of security while also limiting customization options for users. Apple's stringent app review process and hardware–software integration create a more secure environment.
- **Windows**, being a popular OS for both desktop and mobile devices, has encountered security challenges due to its widespread use. Microsoft has implemented various security measures, such as Windows Defender, to address these challenges and enhance overall system security.

- **Symbian OS** faced challenges such as limited processing power and memory, requiring efficient resource management. It introduced advanced features such as multitasking and a customizable user interface. However, it also faced security threats such as malware attacks, unauthorized access, and app permission vulnerabilities. Efforts to enhance security include encryption, access control mechanisms, and security policies. Despite being phased out, Symbian remains significant in understanding mobile OS evolution and ongoing efforts to ensure robust and secure platforms.
- **Tizen and Sailfish**, although not as widely adopted as Android and iOS, offer alternative options with their unique features and design philosophies. Their smaller market share may pose challenges in terms of dedicated research, resources, and app ecosystems, which can impact both innovation and security measures.
- **Ubuntu Touch**, known for its convergence capability, strives to strike a balance between security and openness. It benefits from the security measures inherent in Linux-based systems and actively addresses vulnerabilities through regular updates and community support.
- **KaiOS**, primarily used in feature phones, presents its own set of challenges and innovations. With a restricted app ecosystem and a focus on simplicity, KaiOS aims to provide a secure environment for users while catering to the needs of less resource-intensive devices.
- **Sirin** emphasizes enhanced privacy and security, catering to users seeking heightened protection. Its focus on secure communication and encryption demonstrates innovation in providing a robust security framework.
- **HarmonyOS**, a relatively new entrant, aims to offer a seamless and secure experience across different device types. Its distributed architecture and emphasis on privacy and security show promising innovations for the future of operating systems.

Table 8.1 provides a comparative analysis of various mobile OSs such as Android, iOS, Windows, Symbian, Tizen, Sailfish, Ubuntu Touch, KaiOS, Sirin, and HarmonyOS.

The table highlights the challenges faced by each OS such as app security, fragmentation, restricted customization, and limited resources. It also outlines the security implications and measures taken to address them, such as regular updates, app vetting, sandboxing, and hardware integration.

Additionally, key findings are summarized, emphasizing the trade-offs between innovation and security, the importance of balancing openness with robust security practices, and the potential for unique design philosophies and convergence capabilities.

We also discussed mobile application security, which significantly impacts users' lives and can lead to various security threats. For instance, attackers sometimes try to exploit mobile device vulnerabilities to steal personal information and track users. Mobile application threats can occur in various layers, including application, transport, network, data link, and physical layers, resulting in data collision, denial

8 Conclusion

Table 8.1 Comparative analysis: challenges, security implications, and key findings of mobile OS

OS	Challenges	Security implication	Key findings
Android	Diverse ecosystem, fragmentation, app security	Regular updates, sandboxing, app vetting	Open-source nature: innovation vs. security risks
iPhone	Restricted customization, controlled app ecosystem	App review process, hardware in integration	Tightly controlled ecosystem: enhanced security, limited customization
Windows	Widespread usage, vulnerability to targeted attacks	Windows defender, patch management	Addressing vulnerabilities, strengthening security
Symbian	Limited processing power and memory	Efficient resource management and optimization required	Effective resource management was crucial for smooth performance
Tizen	Niche market, limited resources, app ecosystem	Community support, regular updates	Balancing security and openness, convergence capability
Sailfish	Limited market share, app ecosystem, research focus	Strengthened security measures, collaboration	Unique design philosophy, alternative with potential for innovation
Ubuntu Touch	Limited adoption, resource availability	Linux-based security, regular updates	Convergence capability, inherent security of Linux
KaiOS	Feature phone platform, restricted app ecosystem	App vetting, simplicity in design	Secure experience for less resource-intensive devices
Sirin	Enhanced privacy, limited market share	Secure communication, encryption	Focus on privacy and security for user protection
HarmonyOS	Emerging OS, distributed architecture	Privacy features, cross-device security	Seamless experience, emphasis on privacy and security

of service, spoofing, sinkhole, flooding, replay attacks, and jamming. Importantly, mobile application vulnerabilities can result from design flaws, programming errors, and trade-offs between security and functionality.

Organizations should employ a software assurance process for mobile apps to mitigate security risks. Data confidentiality, authorization, access control, identity authentication, and privacy protection are vital for mobile application security. To ensure mobile application security, we presented the importance of information sensitivity, app vulnerabilities, semantic security, infrastructure-related security, and privacy awareness.

Additionally, we highlighted the need for smartphone users to prioritize device security in the face of increasing cybersecurity crimes. We offered a comprehensive list of best practices to enhance mobile security, including choosing the right mobile operating system, implementing user authentication methods, regularly updating devices and apps, avoiding public Wi-Fi, and turning off unused features such as Bluetooth and Wi-Fi. We also emphasized the importance of utilizing password managers, VPNs, antivirus software, encryption, and being cautious of phishing scams. Finally, we suggested implementing remote lock and data wipe policies, avoiding jailbreaking devices, utilizing mobile device management, conducting regular security audits, and educating employees on mobile security practices.

By following these recommendations, smartphone users can protect their devices and data from potential threats, ensuring higher security and reducing the risk of data breaches or unauthorized access. Prioritizing mobile security and implementing these best practices can help users stay one step ahead of cybercriminals and maintain the privacy and integrity of their personal and business information.

Finally, the key takeaways from this book are as follows:

- Each OS has its own challenges, ranging from maintaining security across diverse ecosystems to managing market share and resources.
- Every OS strives to strike a balance between innovation and security. Innovations in user experience, convergence, privacy features, and hardware–software integration contribute to the evolution of OS while addressing security concerns.
- Different OS employs varying security measures such as the app vetting processes, sandboxing, hardware integration, and secure communication protocols. These measures protect user data, mitigate vulnerabilities, and provide secure computing environments.
- Some OS may have limited documentation, restricted access to proprietary information, or fewer research studies dedicated to them. This highlights the need for ongoing research and collaboration to improve understanding and enhance security practices.
- Every OS must adapt to the ever-changing cybersecurity landscape, promptly addressing emerging threats and vulnerabilities. Regular updates, patch management, and community support are crucial in maintaining secure operating environments.
- Mobile application security is of the utmost importance due to the potential impact on users' lives and the modern prevalence of security threats. Attackers exploit vulnerabilities in mobile devices, compromising personal information and tracking users.
- To mitigate risks, organizations should prioritize data confidentiality, authorization, access control, identity authentication, and privacy protection and implement a software assurance process for mobile apps.
- In light of the growing threat of cybersecurity-related crimes, it is essential for smartphone users to prioritize device security regardless of the operating system they use. With this goal in mind, we provided a comprehensive list of best practices to strengthen mobile security.

References

1. Amin, M., Shah, B., Sharif, A., Ali, T., Kim, K. I., & Anwar, S. (2022). Android malware detection through generative adversarial networks. *Transactions on Emerging Telecommunications Technologies, 33*(2), e3675.
2. Kiss, N., Lalande, J. F., Leslous, M., & Tong, V. V. T. (2016). Kharon dataset: Android malware under a microscope. In *The LASER Workshop: Learning from Authoritative Security Experiment Results (LASER 2016)* (pp. 1–12).
3. Lookout. (2019). *Phishing sites distributing IOS & android surveillanceware*. Lookout Cloud & Endpoint Security. https://www.lookout.com/blog/esurv-research
4. Felt, A. P., et al. (2011). A survey of mobile malware in the wild. In *Proceedings of the 1st ACM workshop on Security and privacy in smartphones and mobile devices*.
5. Grønli, T. M., Hansen, J., Ghinea, G., & Younas, M. (2014, May). Mobile application platform heterogeneity: Android vs. Windows Phone vs. iOS vs. Firefox OS. In *2014 IEEE 28th international conference on advanced information networking and applications* (pp. 635–641). IEEE.
6. Ahvanooey, M. T., Li, Q., Rabbani, M., & Rajput, A. R. (2020). *A survey on smartphone security: Software vulnerabilities, malware, and attacks*. arXiv preprint arXiv:2001.09406.
7. Zahran, O. (2020). KaiOS: The most important operating system. *The Startup*. Accessed 2023, from https://medium.com/swlh/kaios-the-most-important-operating-system-2d92644959a4
8. SIRIN OS, Our Ecosystem. SIRIN LABS. (2023). Accessed 2023, from https://sirinlabs.com/sirin-os/
9. Ozhayta, A. M. (2020). *HarmonyOS*. HUAWEI Developers. Accessed 2023, from https://medium.com/huawei-developers/harmonyos-4bfe31c99be7
10. Li, B., Reshetova, E., Aura, T. (2010). *Symbian OS platform security model. For a complete list of all USENIX & USENIX co-sponsored events*. see http://www.usenix.org/events
11. Balapour, A., Nikkhah, H. R., & Sabherwal, R. (2020). Mobile application security: Role of perceived privacy as the predictor of security perceptions. *International Journal of Information Management, 52*, 102063.
12. Osman, T., Mannan, M., Hengartner, U., & Youssef, A. (2020). Securing applications against side-channel attacks through resource access veto. *Digital Threats: Research and Practice, 1*(4), 1–29.

GPSR Compliance

The European Union's (EU) General Product Safety Regulation (GPSR) is a set of rules that requires consumer products to be safe and our obligations to ensure this.

If you have any concerns about our products, you can contact us on

ProductSafety@springernature.com

In case Publisher is established outside the EU, the EU authorized representative is:

Springer Nature Customer Service Center GmbH
Europaplatz 3
69115 Heidelberg, Germany

www.ingramcontent.com/pod-product-compliance
Lightning Source LLC
Chambersburg PA
CBHW071741030725
29116CB00003B/115